Predictive Analytics with Microsoft Azure Machine Learning

Second Edition

Roger Barga

Valentine Fontama

Wee Hyong Tok

Apress®

Predictive Analytics with Microsoft Azure Machine Learning

ISBN-13 (pbk): 978-1-4842-1201-1

ISBN-13 (electronic): 978-1-4842-1200-4

Managing Director: Welmoed Spahr
Lead Editor: James DeWolf
Development Editor: Douglas Pundick
Technical Reviewers: Luis Cabrera-Cordon, Jacob Spoelstra, Hang Zhang, and Yan Zhang

Coordinating Editor: Melissa Maldonado
Copy Editor: Mary Behr
Compositor: SPi Global
Indexer: SPi Global
Artist: SPi Global

Distributed to the book trade worldwide by Springer Science+Business Media New York, 233 Spring Street, 6th Floor, New York, NY 10013. Phone 1-800-SPRINGER, fax (201) 348-4505, e-mail orders-ny@springer-sbm.com, or visit www.springeronline.com. Apress Media, LLC is a California LLC and the sole member (owner) is Springer Science + Business Media Finance Inc (SSBM Finance Inc). SSBM Finance Inc is a **Delaware** corporation.

For information on translations, please e-mail rights@apress.com, or visit www.apress.com.

Apress and friends of ED books may be purchased in bulk for academic, corporate, or promotional use. eBook versions and licenses are also available for most titles. For more information, reference our Special Bulk Sales–eBook Licensing web page at www.apress.com/bulk-sales.

Any source code or other supplementary material referenced by the author in this text is available to readers at www.apress.com. For detailed information about how to locate your book's source code, go to www.apress.com/source-code/.

Contents at a Glance

Contents

About the Authors

Roger Barga is a General Manager and Director of Development at Amazon Web Services. Prior to joining Amazon, Roger was Group Program Manager for the Cloud Machine Learning group in the Cloud & Enterprise division at Microsoft, where his team was responsible for product management of the Azure Machine Learning service. Roger joined Microsoft in 1997 as a Researcher in the Database Group of Microsoft Research, where he directed both systems research and product development efforts in database, workflow, and stream processing systems. He has developed ideas from basic research, through proof of concept prototypes, to incubation efforts in product groups. Prior to joining Microsoft, Roger was a Research Scientist in the Machine Learning Group at the Pacific Northwest National Laboratory where he built and deployed machine learning-based solutions. Roger is also an Affiliate Professor at the University of Washington, where he is a lecturer in the Data Science and Machine Learning programs.

Roger holds a PhD in Computer Science, a M.Sc. in Computer Science with an emphasis on Machine Learning, and a B.Sc. in Mathematics and Computing Science. He has published over 90 peer-reviewed technical papers and book chapters, collaborated with 214 co-authors from 1991 to 2013, with over 700 citations by 1,084 authors.

Valentine Fontama is a Data Scientist Manager in the Cloud & Enterprise Analytics and Insights team at Microsoft. Val has over 18 years of experience in data science and business. Following a PhD in Artificial Neural Networks, he applied data mining in the environmental science and credit industries. Before Microsoft, Val was a New Technology Consultant at Equifax in London where he pioneered the application of data mining to risk assessment and marketing in the consumer credit industry. He is currently an Affiliate Professor of Data Science at the University of Washington.

In his prior role at Microsoft, Val was a Principal Data Scientist in the Data and Decision Sciences Group (DDSG) at Microsoft, where he led external consulting engagements with Microsoft's customers, including ThyssenKrupp and Dell. Before that he was a Senior Product Marketing Manager responsible for big data and predictive analytics in cloud and enterprise marketing. In this role, he led product management for Microsoft Azure Machine Learning; HDInsight, the first Hadoop service from Microsoft; Parallel Data Warehouse, Microsoft's first data warehouse appliance; and three releases of Fast Track Data Warehouse.

Val holds an M.B.A. in Strategic Management and Marketing from Wharton Business School, a Ph.D. in Neural Networks, a M.Sc. in Computing, and a B.Sc. in Mathematics and Electronics (with First Class Honors). He co-authored the book *Introducing Microsoft Azure HDInsight*, and has published 11 academic papers with 152 citations by over 227 authors.

Wee-Hyong Tok is a Senior Program Manager of the Information Management and Machine Learning (IMML) team in the Cloud and Enterprise group at Microsoft Corp. Wee-Hyong brings decades of database systems experience, spanning industry and academia.

Prior to pursuing his PhD, Wee-Hyong was a System Analyst at a large telecommunication company in Singapore. Wee-Hyong was a SQL Server Most Valuable Professional (MVP), specializing in business intelligence and data mining. He was responsible for spearheading data mining boot camps in Southeast Asia, with a goal of empowering IT professionals with the knowledge and skills to use analytics in their organization to turn raw data into insights.

He joined Microsoft and worked on the SQL Server team, and is responsible for shaping the SSIS Server, bringing it from concept to release in SQL Server 2012.

Wee Hyong holds a Ph.D. in Computer Science, M.Sc. in Computing, and a B.Sc. (First Class Honors) in Computer Science, from the National University of Singapore. He has published 21 peer reviewed academic papers and journals. He is a co-author of the following books: *Predictive Analytics with Microsoft Azure Machine Learning, Introducing Microsoft Azure HDInsight, and Microsoft SQL Server 2012 Integration Services.*

About the Technical Reviewers

Luis Cabrera-Cordon is a Program Manager in the Azure Machine Learning Team, where he focuses on the Azure Machine Learning APIs and the new Machine Learning Marketplace. He is passionate about interaction design and creating software development platforms that are accessible and exciting to use. Luis has worked at Microsoft for over 12; before Azure Machine Learning, he was the Program Manager Lead in charge of the Bing Maps development platform and the PM in charge of the Microsoft Surface developer platform (remember the big Surface?). In a previous life, he was a developer on the Windows Mobile team, working on the first managed APIs that shipped in Windows Mobile. Outside of work, Luis enjoys spending time with his family in the Pacific Northwest. He holds a Masters in Software Engineering from the University of Washington.

Jacob Spoelstra is a Director of Data Science at Microsoft, where he leads a group in the Azure Machine Learning organization responsible for both building end-to-end predictive analytics solutions for internal clients and helping partners adopt the platform. He has more than two decades experience in machine learning and predictive analytics, focusing in particular on neural networks.

Prior to Microsoft, Jacob was the global head of R&D at Opera Solutions. Under his watch, the R&D team developed key Opera innovations, including a patented adaptive auto auction pricing solution utilizing Kalman filters, and collaborative filtering techniques to find statistical anomalies in medical bills as a way to detect revenue leakage. He headed up the Opera Solutions team that, as part of "The Ensemble," ended up "first equals" in the prestigious Netflix data mining competition, beating out over 41,000 other entrants.

Jacob has held analytics leadership positions at FICO, SAS, ID Analytics, and boutique consulting company BasePoint. He holds BS and MS degrees in Electrical Engineering from the University of Pretoria, and a PhD in Computer Science from the University of Southern California.

He and his wife, Tanya, have two boys, aged 10 and 12. They enjoy camping, hiking, and snow sports. Jacob is a private pilot and is constantly looking for excuses to go flying.

Dr. Hang Zhang joined Microsoft in May 2014 as a Senior Data Scientist, Cloud Machine Learning Data Science. Before joining Microsoft, Hang was a Staff Data Scientist at WalmartLabs leading a team building internal tools for search analytics and business intelligence. He worked for two years in Opera Solutions as a Senior Scientist focusing on machine learning and data science between 2011 and 2013. Before that, Hang worked at Arizona State University for four years in the area of neuro-informatics. Hang holds a Ph.D. degree in Industrial and Systems Engineering, and a M.S. degree in Statistics from Rutgers, The State University of New Jersey.

Dr. Yan Zhang is a senior data scientist in Microsoft Cloud & Enterprise Azure Machine Learning product team. She builds predictive models and generalized data driven solutions on the Cloud machine learning platform. Her recent research includes predictive maintenance in IoT applications, customer segmentation, and text mining. Dr. Zhang received her Ph.D. in data mining. Before joining Microsoft, she was a research faculty at Syracuse University, USA.

Acknowledgments

I would like to express my gratitude to the many people in the Azure ML team at Microsoft who saw us through this book; to all those who provided support, read, offered comments, and assisted in the editing, and proofreading. I wish to thank my coauthors, Val and Wee-Hyong, for their drive and perseverance, which was key to completing this book, and to our publisher Apress, especially Melissa Maldonado and James T. DeWolf, for making this all possible. Above all I want to thank my wife, Terre, and my daughters Amelie and Jolie, who supported and encouraged me in spite of all the time it took me away from them.

—Roger Barga

I would like to thank my co-authors, Roger and Wee-Hyong, for their deep collaboration on this project. I am grateful to all our reviewers and editors whose input was critical to the success of the book. Special thanks to my wife, Veronica, and loving kids, Engu, Chembe, and Nayah, for their support and encouragement through two editions of this book.

—Valentine Fontama

I would like to thank my coauthors, Roger and Val, for the great camaraderie on this journey to deliver the second edition of this book. I deeply appreciate the reviews by the team of data scientists from the machine learning team, and the feedback from readers all over the world after we shipped the first edition. This feedback helped us to improve this book tremendously. I'd also like to thank the Apress team for working with us to shape the second edition. Special thanks to my family, Juliet, Nathaniel, Siak-Eng, and Hwee-Tiang, for their love, support, and patience.

—Wee-Hyong Tok

Foreword

Few people appreciate the enormous potential of machine learning (ML) in enterprise applications. I was lucky enough to get a taste of its potential benefits just a few months into my first job. It was 1995, and credit card issuers were beginning to adopt neural network models to detect credit card fraud in real time. When a credit card is used, transaction data from the point of sale system is sent to the card issuer's credit authorization system where a neural network scores for the probability of fraud. If the probability is high, the transaction is declined in real time. I was a scientist building such models, and one of my first model deliveries was for a South American bank. When the model was deployed, the bank identified over a million dollars of previously undetected fraud on the very first day. This was a big eye-opener. In the years since, I have seen ML deliver huge value in diverse applications such as demand forecasting, failure and anomaly detection, ad targeting, online recommendations, and virtual assistants like Cortana. By embedding ML into their enterprise systems, organizations can improve customer experience, reduce the risk of systemic failures, grow revenue, and realize significant cost savings.

However, building ML systems is slow, time-consuming, and error prone. Even though we are able to analyze very large data sets these days and deploy at very high transaction rates, the following bottlenecks remain:

- ML system development requires deep expertise. Even though the core principles of ML are now accessible to a wider audience, talented data scientists are as hard to hire today as they were two decades ago.
- Practitioners are forced to use a variety of tools to collect, clean, merge, and analyze data. These tools have a steep learning curve and are not integrated. Commercial ML software is expensive to deploy and maintain.
- Building and verifying models requires considerable experimentation. Data scientists often find themselves limited by compute and storage because they need to run a large number of experiments that generate considerable new data.
- Software tools do not support scalable experimentation or methods for organizing experiment runs. The act of collaborating with a team on experiments and sharing derived variables, scripts, etc. is manual and ad-hoc without tools support. Evaluating and debugging statistical models remains a challenge.

Data scientists work around these limitations by writing custom programs and by doing undifferentiated heavy lifting as they perform their ML experiments. But it gets harder in the deployment phase. Deploying ML models in a mission-critical business process such as real-time fraud prevention or ad targeting requires sophisticated engineering. The following needs must be met:

- Typically, ML models that have been developed offline now have to be reimplemented in a language such as C++, C#, or Java.
- The transaction data pipelines have to be plumbed. Data transformations and variables used in the offline models have to be recoded and compiled.
- These reimplementations inevitably introduce bugs, requiring verification that the models work as originally designed.
- A custom container for the model has to be built, with appropriate monitors, metrics, and logging.
- Advanced deployments require A/B testing frameworks to evaluate alternative models side by side. One needs mechanisms to switch models in or out, preferably without recompiling and deploying the entire application.
- One has to validate that the candidate production model works as originally designed through statistical tests.
- The automated decisions made by the system and the business outcomes have to be logged for refining the ML models and for monitoring.
- The service has to be designed for high availability, disaster recovery, and geo-proximity to end points.
- When the service has to be scaled to meet higher transaction rates and/or low latency, more work is required to provision new hardware, deploy the service to new machines, and scale out.

These are time-consuming and engineering-intensive steps, expensive in terms of both infrastructure and manpower. The end-to-end engineering and maintenance of a production ML application requires a highly skilled team that few organizations can build and sustain.

Microsoft Azure ML was designed to solve these problems.

- It's a fully managed cloud service with no software to install, no hardware to manage, and no OS versions or development environments to grapple with.
- Armed with nothing but a browser, data scientists can log on to Azure and start developing ML models from any location, from any device. They can host a practically unlimited number of files on Azure storage.

- ML Studio, an integrated development environment for ML, lets you set up experiments as simple data flow graphs, with an easy-to-use drag, drop, and connect paradigm. Data scientists can avoid programming for a large number of common tasks, allowing them to focus on experiment design and iteration.
- Many sample experiments are provided to make it easy to get started.
- A collection of best-of-breed algorithms developed by Microsoft Research is built in, as is support for custom R code. Over 350 open source R packages can be used securely within Azure ML.
- Data flow graphs can have several parallel paths that automatically run in parallel, allowing scientists to execute complex experiments and make side-by-side comparisons without the usual computational constraints.
- Experiments are readily sharable, so others can pick up on your work and continue where you left off.

Azure ML also makes it simple to create production deployments at scale in the cloud. Pretrained ML models can be incorporated into a scoring workflow and, with a few clicks, a new cloud-hosted REST API can be created. This REST API has been engineered to respond with low latency. No reimplementation or porting is required, which is a key benefit over traditional data analytics software. Data from anywhere on the Internet (laptops, websites, mobile devices, wearables, and connected machines) can be sent to the newly created API to get back predictions. For example, a data scientist can create a fraud detection API that takes transaction information as input and returns a low/medium/high risk indicator as output. Such an API would then be "live" on the cloud, ready to accept calls from any software that a developer chooses to call it from. The API backend scales elastically, so that when transaction rates spike, the Azure ML service can automatically handle the load. There are virtually no limits on the number of ML APIs that a data scientist can create and deploy–and all this without any dependency on engineering. For engineering and IT, it becomes simple to integrate a new ML model using those REST APIs, and testing multiple models side-by-side before deployment becomes easy, allowing dramatically better agility at low cost. Azure provides mechanisms to scale and manage APIs in production, including mechanisms to measure availability, latency, and performance. Building robust, highly available, reliable ML systems and managing the production deployment is therefore dramatically faster, cheaper, and easier for the enterprise, with huge business benefits.

We believe Azure ML is a game changer. It makes the incredible potential of ML accessible both to startups and large enterprises. Startups are now able to use the same capabilities that were previously available to only the most sophisticated businesses. Larger enterprises are able to unleash the latent value in their big data to generate significantly more revenue and efficiencies. Above all, the speed of iteration and experimentation that is now possible will allow for rapid innovation and pave the way for intelligence in cloud-connected devices all around us.

When I started my career in 1995, it took a large organization to build and deploy credit card fraud detection systems. With tools like Azure ML and the power of the cloud, a single talented data scientist can accomplish the same feat. The authors of this book, who have long experience with data science, have designed it to help you get started on this wonderful journey with Azure ML.

—Joseph Sirosh
Corporate Vice President,
Machine Learning, Microsoft Corporation

Introduction

Data science and machine learning are in high demand, as customers are increasingly looking for ways to glean insights from their data. More customers now realize that business intelligence is not enough, as the volume, speed, and complexity of data now defy traditional analytics tools. While business intelligence addresses descriptive and diagnostic analysis, data science unlocks new opportunities through predictive and prescriptive analysis.

This book provides an overview of data science and an in-depth view of Microsoft Azure Machine Learning, which is part of the Cortana Analytics Suite. Cortana Analytics Suite is a fully managed big data and advanced analytics suite that helps organizations transform data into intelligent action. This book provides a structured approach to data science and practical guidance for solving real-world business problems such as buyer propensity modeling, customer churn analysis, predictive maintenance, and product recommendation. The simplicity of the Azure Machine Learning service from Microsoft will help to take data science and machine learning to a much broader audience than existing products in this space. Learn how you can quickly build and deploy sophisticated predictive models as machine learning web services with the new Azure Machine Learning service from Microsoft.

Who Should Read This Book?

This book is for budding data scientists, business analysts, BI professionals, and developers. The reader needs to have basic skills in statistics and data analysis. That said, they do not need to be data scientists nor have deep data mining skills to benefit from this book.

What You Will Learn

This book will provide the following:

- A deep background in data science, and how to solve a business data science problem using a structured approach and best practices
- How to use the Microsoft Azure Machine Learning service to effectively build and deploy predictive models as machine learning web services
- Practical examples that show how to solve typical predictive analytics problems such as propensity modeling, churn analysis, and product recommendation

At the end of the book, you will have gained essential skills in basic data science, the data mining process, and a clear understanding of the new Microsoft Azure Machine Learning service. You'll also have the framework to solve practical business problems with machine learning.

PART I

Introducing Data Science and Microsoft Azure Machine Learning

CHAPTER 1

Introduction to Data Science

So what is data science and why is it so topical? Is it just another fad that will fade away after the hype? We will start with a simple introduction to data science, defining what it is, why it matters, and why it matters now. This chapter will highlight the data science process with guidelines and best practices. It will introduce some of the most commonly used techniques and algorithms in data science. And it will explore ensemble models, a key technology on the cutting edge of data science.

What is Data Science?

Data science is the practice of obtaining useful insights from data. Although it also applies to small data, data science is particularly important for big data, as we now collect petabytes of structured and unstructured data from many sources inside and outside an organization. As a result, we are now data rich but information poor. Data science provides powerful processes and techniques for gleaning actionable information from this sea of data. Data science draws from several disciplines including statistics, mathematics, operations research, signal processing, linguistics, database and storage, programming, machine learning, and scientific computing. Figure 1-1 illustrates the most common disciplines of data science. Although the term *data science* is new in business, it has been around since 1960 when it was first used by Peter Naur to refer to data processing methods in computer science. Since the late 1990s notable statisticians such as C.F. Jeff Wu and William S. Cleveland have also used the term data science, a discipline they view as the same as or an extension of statistics.

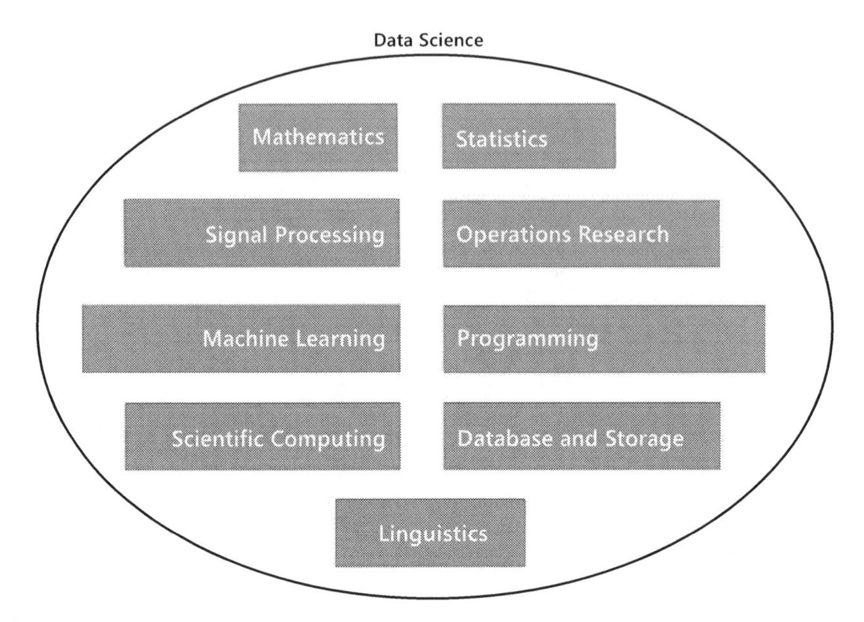

Figure 1-1. *Highlighting the main academic disciplines that constitute data science*

Practitioners of data science are data scientists, whose skills span statistics, mathematics, operations research, signal processing, linguistics, database and storage, programming, machine learning, and scientific computing. In addition, to be effective, data scientists also need good communication and data visualization skills. Domain knowledge is also important to deliver meaningful results fast. This breadth of skills is very hard to find in one person, which is why data science is a team sport, not an individual effort. To be effective, one needs to hire a team with complementary data science skills.

Analytics Spectrum

According to Gartner, all the analytics we do can be classified into one of four categories: descriptive, diagnostic, predictive, and prescriptive analysis. Descriptive analysis typically helps to describe a situation and can help to answer questions like *What happened?, Who are my customers*?, etc. Diagnostic analysis help you understand why things happened and can answer questions like *Why did it happen?* Predictive analysis is forward-looking and can answer questions such as *What will happen in the future?* As the name suggests, prescriptive analysis is much more prescriptive and helps answer questions like *What should we do?, What is the best route to my destination?*, or *How should I allocate my investments?*

Figure 1-2 illustrates the full analytics spectrum. It also shows the degree of sophistication in this diagram.

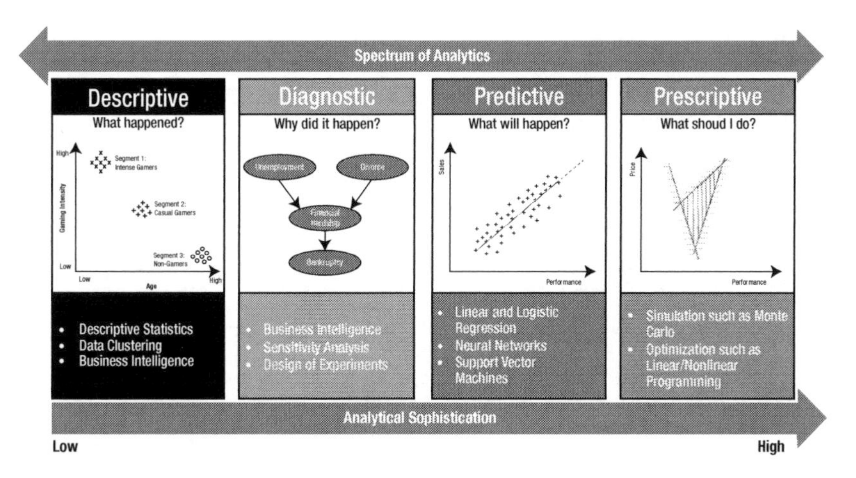

Figure 1-2. *Spectrum of all data analysis*

Descriptive Analysis

Descriptive analysis is used to explain what is happening in a given situation. This class of analysis typically involves human intervention and can be used to answer questions like *What happened?, Who are my customers?, How many types of users do we have?*, etc. Common techniques used for this include descriptive statistics with charts, histograms, box and whisker plots, or data clustering. You'll explore these techniques later in this chapter.

Diagnostic Analysis

Diagnostic analysis helps you understand why certain things happened and what are the key drivers. For example, a wireless provider would use this to answer questions such as *Why are dropped calls increasing?* or *Why are we losing more customers every month?* A customer diagnostic analysis can be done with techniques such as clustering, classification, decision trees, or content analysis. These techniques are available in statistics, data mining, and machine learning. It should be noted that business intelligence is also used for diagnostic analysis.

Predictive Analysis

Predictive analysis helps you predict what will happen in the future. It is used to predict the probability of an uncertain outcome. For example, it can be used to predict if a credit card transaction is fraudulent, or if a given customer is likely to upgrade to a premium phone plan. Statistics and machine learning offer great techniques for prediction. This includes techniques such as neural networks, decision trees, random forests, boosted decision trees, Monte Carlo simulation, and regression.

Prescriptive Analysis

Prescriptive analysis will suggest the best course of action to take to optimize your business outcomes. Typically, prescriptive analysis combines a predictive model with business rules (such as declining a transaction if the probability of fraud is above a given threshold). For example, it can suggest the best phone plan to offer a given customer, or based on optimization, can propose the best route for your delivery trucks. Prescriptive analysis is very useful in scenarios such as channel optimization, portfolio optimization, or traffic optimization to find the best route given current traffic conditions. Techniques such as decision trees, linear and non-linear programming, Monte Carlo simulation, or game theory from statistics and data mining can be used to do prescriptive analysis. See Figure 1-3.

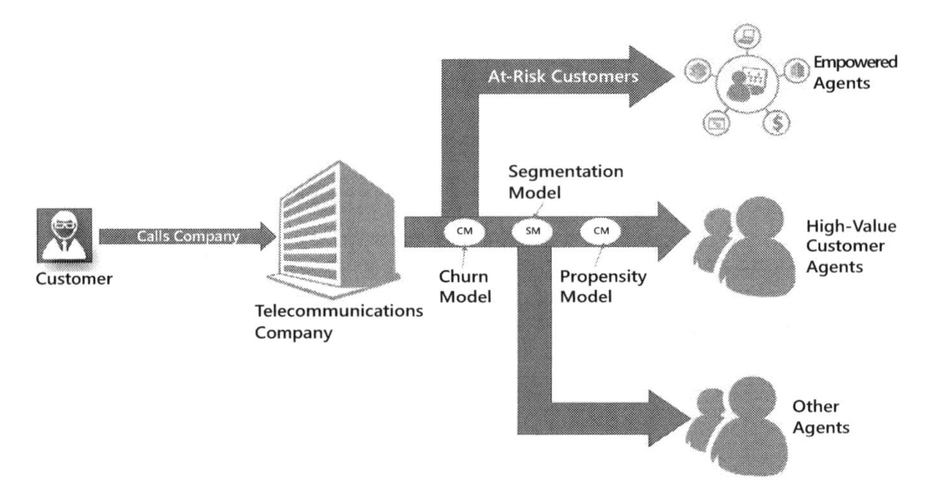

Figure 1-3. *A smart telco using prescriptive analytics*

The analytical sophistication increases from descriptive to prescriptive analytics. In many ways, prescriptive analytics is the nirvana of analytics and is often used by the most analytically sophisticated organizations. Imagine a smart telecommunications company that has embedded analytical models in its business workflow systems. It has the following analytical models embedded in its customer call center system:

- **A Customer Churn Model**: This is a predictive model that predicts the probability of customer attrition. In other words, it predicts the likelihood that the customer calling the call center will ultimately defect to the competition.
- **A Customer Segmentation Model**: This puts customers into distinct behavioral segments for marketing purposes.
- **A Customer Propensity Model**: This model predicts the customer's propensity to respond to each of the marketing offers, such as upgrades to premium plans.

When a customer calls, the call center system identifies him or her in real time from their cell phone number. Then the call center system scores the customer using these three models. If the customer scores high on the customer churn model, it means they are very likely to defect to the competitor. In that case, the telecommunications company will immediately route the customer to a group of call center agents who are empowered to make attractive offers to prevent attrition. Otherwise, if the segmentation model scores the customer as a profitable customer, he/she is routed to a special concierge service with shorter wait lines and the best customer service. If the propensity model scores the customer high for upgrades, the call agent is alerted and will try to upsell the customer with attractive upgrades. The beauty of this solution is that all of the models are baked into the telecommunication company's business workflow, which enables their agents to make smart decisions that improve profitability and customer satisfaction. This is illustrated in Figure 1-3.

Why Does It Matter and Why Now?

Data science offers organizations a real opportunity to make smarter and timely decisions based on all the data they collect. With the right tools, data science offers you new and actionable insights not only from your own data, but also from the growing sources of data outside your organization, such as weather data, customer demographic data, consumer credit data from the credit bureaus, and data from social media sites such as Facebook, Twitter, Instagram, etc. Here are a few reasons why data science is now critical for business success.

Data as a Competitive Asset

Data is now a critical asset that offers a competitive advantage to smart organizations that use it correctly for decision making. McKinsey and Gartner agree on this: in a recent paper McKinsey suggests that companies that use data and business analytics to make decisions are more productive and deliver a higher return on equity than those who don't. In a similar vein, Gartner posits that organizations that invest in a modern data infrastructure will outperform their peers by up to 20%. Big data offers organizations the opportunity to combine valuable data across silos to glean new insights that drive smarter decisions.

> *"Companies that use data and business analytics to guide decision making are more productive and experience higher returns on equity than competitors that don't."*
>
> —Brad Brown et al., McKinsey Global Institute, 2011

> *"By 2015, organizations integrating high-value, diverse, new information types and sources into a coherent information management infrastructure will outperform their industry peers financially by more than 20%."*
>
> —Regina Casonato et al., Gartner[1]

Increased Customer Demand

Business intelligence has been the key type of analytics used by most organizations in the last few decades. However, with the emergence of big data, more customers are now eager to use predictive analytics to improve marketing and business planning. Traditional BI gives a good rear view analysis of their business, but does not help with any forward-looking questions that include forecasting or prediction.

The past two years have seen a surge of demand from customers for predictive analytics as they seek more powerful analytical techniques to uncover value from the troves of data they store on their businesses. In our combined experience, we have not seen as much demand for data science from customers as we did in the last two years alone!

Increased Awareness of Data Mining Technologies

Today a subset of data mining and machine learning algorithms are now more widely understood since they have been tried and tested by early adopters such as Netflix and Amazon, who actively use them in their recommendation engines. While most customers do not fully understand details of the machine learning algorithms used, their application in Netflix movie recommendations or recommendation engines at online stores are very salient. Similarly, many customers are now aware of the targeted ads that are now heavily used by most sophisticated online vendors. So while many customers may not know details of the algorithms used, they now increasingly understand their business value.

Access to More Data

Digital data has exploded in the last few years and shows no signs of abating. Most industry pundits now agree that we are collecting more data than ever before. According to IDC, the digital universe will grow to 35 zetabyes (i.e. 35 trillion terabytes) globally by 2020. Others posit that the world's data is now growing by up to 10 times every 5 years, which is astounding. In a recent study, McKinsey Consulting also found that in 15 of the 17 US economic sectors, companies with over 1,000 employees store, on average, over 235 terabytes of data-which is more than the data stored by the US Library of Congress! This data explosion is driven by the rise of new data sources such as social media, cell phones, smart sensors, and dramatic gains in the computer industry. The rise of Internet of Things (IoT) only exacerbates this trend as more data is being generated than ever before by sensors. According to Cisco, there will be up to 50 billion connected devices by 2020!

The large volumes of data being collected also enable you to build more accurate predictive models. We know from statistics that the confidence interval (also known as the margin of error) has an inverse relationship with the sample size. So the larger your sample size, the smaller the margin of error. This in turn increases the accuracy of predictions from your model.

Faster and Cheaper Processing Power

We now have far more computing power at our disposal than ever before. Moore's Law proposed that computer chip performance would grow exponentially, doubling every 18 months. This trend has been true for most of the history of modern computing. In 2010, the International Technology Roadmap for Semiconductors updated this forecast, predicting that growth would slow down in 2013 when computer densities and counts would double every 3 years instead of 18 months. Despite this, the exponential growth in processor performance has delivered dramatic gains in technology and economic productivity. Today, a smartphone's processor is up to five times more powerful than that of a desktop computer 20 years ago. For instance, the Nokia Lumia 928 has a dual-core 1.5 GHz Qualcomm Snapdragon™ S4 that is at least five times faster than the Intel Pentium P5 CPU released in 1993, which was very popular for personal computers. In the nineties, expensive workstations like the DEC VAX mainframes or the DEC Alpha workstations were required to run advanced, compute-intensive algorithms. It is remarkable that today's smartphone is also five times faster than the powerful DEC Alpha processor from 1994, whose speed was 200-300 MHz! Today you can run the same algorithms on affordable personal workstations with multi-core processors. In addition, you can leverage Hadoop's MapReduce architecture to deploy powerful data mining algorithms on a farm of commodity servers at a much lower cost than ever before. With data science we now have the tools to discover hidden patterns in our data through smart deployment of data mining and machine learning algorithms.

We have also seen dramatic gains in capacity, and an exponential reduction in the price of computer memory. This is illustrated in Figures 1-4 and 1-5, which show the exponential price drop and growth in capacity of computer memory since 1960. Since 1990, the average price per MB of memory has dropped from $59 to a meager 0.49 cents–a 99.2% price reduction! At the same time, the capacity of a memory module has increased from 8MB to a whopping 8GB! As a result, a modest laptop is now more powerful than a high-end workstation from the early nineties.

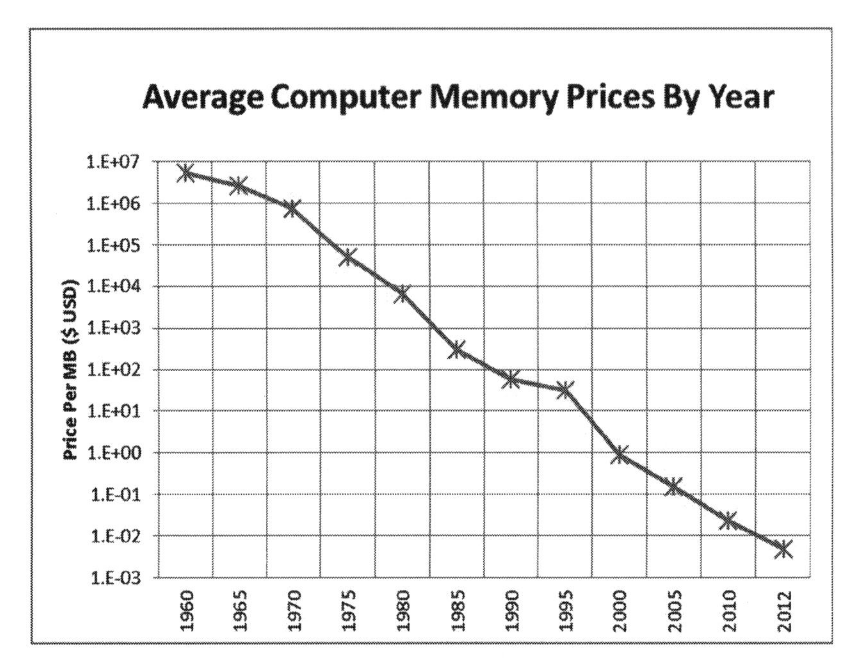

Figure 1-4. *Average computer memory price since 1960*

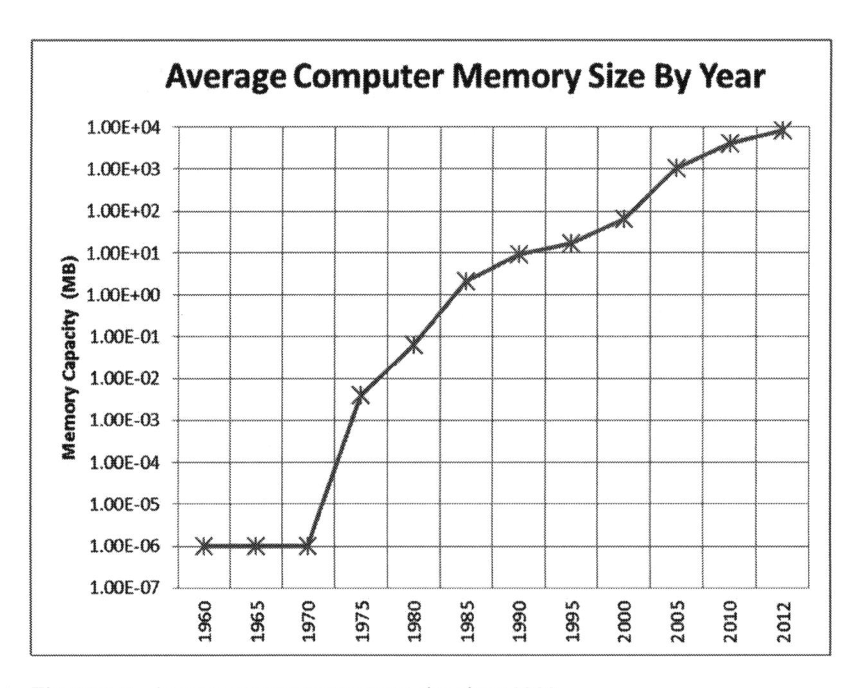

Figure 1-5. *Average computer memory size since 1960*

■ **Note** More information on memory price history is available at John C. McCallum at `www.jcmit.com/mem2012.htm`.

The Data Science Process

A typical data science project follows the five-step process outlined in Figure 1-6. Let's review each of these steps in detail.

1. **Define the business problem**: This is critical as it guides the rest of the project. Before building any models, it is important to work with the project sponsor to identify the specific business problem he or she is trying to solve. Without this, one could spend weeks or months building sophisticated models that solve the wrong problem, leading to wasted effort. A good data science project gleans good insights that drive smarter business decisions. Hence the analysis should serve a business goal. It should not be a hammer in search of a nail! There are formal consulting techniques and frameworks (such as guided discovery workshops and six sigma methodology) used by practitioners to help business stakeholders prioritize and scope their business goals.

2. **Acquire and prepare data**: This step entails two activities. The first is the acquisition of raw data from several source systems including databases, CRM systems, web log files, etc. This may involve ETL (extract, transform, and load) processes, database administrators, and BI personnel. However, the data scientist is intimately involved to ensure the right data is extracted in the right format. Working with the raw data also provides vital context, which is required downstream.

   Second, once the right data is pulled, it is analyzed and prepared for modelling. This involves addressing missing data, outliers in the data, and data transformations. Typically, if a variable has over 40% of missing values, it can be rejected, unless the fact that it is missing (or not) conveys critical information. For example, there might be a strong bias in the demographics of who fills in the optional field of "age" in a survey. For the rest, we need to decide how to deal with missing values; should we impute with the average value, median, or something else? There are several statistical techniques for detecting outliers. With a box and whisker plot, an outlier is a sample (value) greater or smaller than 1.5 times the interquartile range (IQR). The interquartile range is the 75th percentile-25th percentile. We need to decide whether to drop an outlier or not. If it makes sense to keep it, we need to find a useful transformation for the variable. For instance, log transformation is generally useful for transforming incomes.

Correlation analysis, principal component analysis, or factor analysis are useful techniques that show the relationships between the variables. Finally, feature selection is done at this stage to identify the right variables to use in the model in the next step.

This step can be laborious and time-consuming. In fact, in a typical data science project, we spend up to 75 to 80% of time in data acquisition and preparation. That said, this is the vital step that coverts raw data into high quality gems for modelling. The old adage is still true: *garbage in, garbage out*. Investing wisely in data preparation improves the success of your project. Chapter 3 provides more details on the data preparation phase.

3. **Develop the model**: This is the most fun part of the project where we develop the predictive models. In this step, we determine the right algorithm to use for modeling given the business problem and data. For instance, if it is a binary classification problem, we can use logistic regression, decision trees, boosted decision trees, or neural networks. If the final model has to be explainable, this rules out algorithms like boosted decision trees. Model building is an iterative process: we experiment with different models to find the most predictive one. We also validate it with the customer a few times to ensure it meets their needs before exiting this stage.

4. **Deploy the model**: Once built, the final model has to be deployed in production where it will be used to score transactions or by customers to drive real business decisions. Models are deployed in many different ways depending on the customer's environment. In most cases, deploying a model involves implementing the data transformations and predictive algorithm developed by the data scientist in order to integrate with an existing decision management platform. Needless to say, it is a cumbersome process today. Azure Machine Learning dramatically simplifies model deployment by enabling data scientists to deploy their finished models as web services that can be invoked from any application on any platform, including mobile devices.

5. **Monitor model's performance**: Data science does not end with deployment. It is worth noting that every statistical or machine learning model is only an approximation of the real world, and hence is imperfect from the very beginning. When a validated model is tested and deployed in production, it has to be monitored to ensure it is performing as planned. This is critical because any data-driven model has a fixed shelf life. The accuracy of the model degrades with time because fundamentally the data in production will vary over time for a number of reasons (such as the business may launch new products to target a different

demographic). For instance, the wireless carrier we discussed earlier may choose to launch a new phone plan for teenage kids. If they continue to use the same churn and propensity models, they may see a degradation in their models' performance after the launch of this new product. This is because the original dataset used to build the churn and propensity models did not contain significant numbers of teenage customers. With close monitoring of the model in production we can detect when its performance starts to degrade. When its accuracy degrades significantly, it is time to rebuild the model by either re-training it with the latest dataset including production data, or completely rebuilding it with additional datasets. In that case, we return to Step 1 where we revisit the business goals and start all over.

How often should we rebuild a model? The frequency varies by business domain. In a stable business environment where the data does not vary too quickly, models can be rebuilt once every year or two. A good example is retail banking products such as mortgages and car loans. However, in a very dynamic environment where the ambient data changes rapidly, models can be rebuilt daily or weekly. A good case in point is the wireless phone industry, which is fiercely competitive. Churn models need to be retrained every few days since customers are being lured by ever more attractive offers from the competition.

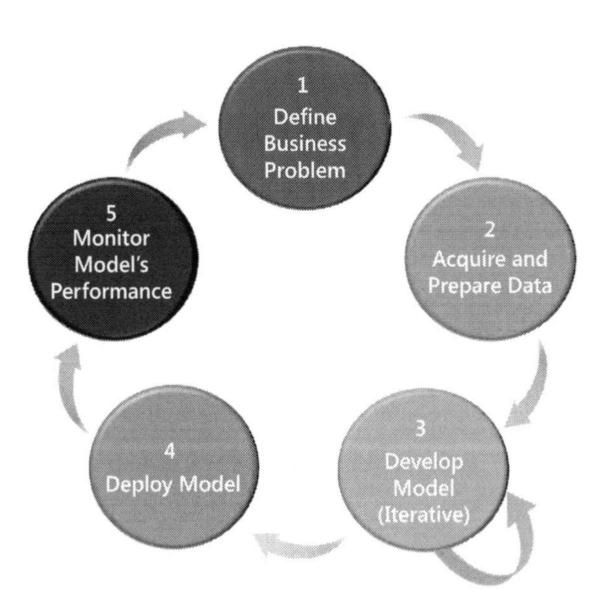

Figure 1-6. *Overview of the data science process*

Common Data Science Techniques

Data science offers a large body of algorithms from its constituent disciplines, namely statistics, mathematics, operations research, signal processing, linguistics, database and storage, programming, machine learning, and scientific computing. We organize these algorithms into the following groups for simplicity:

- Classification
- Clustering
- Regression
- Simulation
- Content Analysis
- Recommenders

Chapter 6 provides more details on some of these algorithms.

Classification Algorithms

Classification algorithms are commonly used to classify people or things into one of many groups. They are also widely used for predictions. For example, to prevent fraud, a card issuer will classify a credit card transaction as either fraudulent or not. The card issuer typically has a large volume of historical credit card transactions and knows the status of each of these transactions. Many of these cases are reported by the legitimate cardholder who does not want to pay for unauthorized charges. So the issuer knows whether each transaction was fraudulent or not. Using this historical data the issuer can now build a model that predicts whether a new credit card transaction is likely to be fraudulent or not. This is a binary classification problem in which all cases fall into one of two classes.

Another classification problem is the customers' propensity to upgrade to a premium phone plan. In this case, the wireless carrier needs to know if a customer will upgrade to a premium plan or not. Using sales and usage data, the carrier can determine which customers upgraded in the past. Hence they can classify all customers into one of two groups: whether they upgraded or not. Since the carrier also has information on demographic and behavioral data on new and existing customers, they can build a model to predict a new customer's probability to upgrade; in other words, the model will group each customer into one of two classes.

Statistics and data mining offer many great tools for classification: this includes logistic regression, which is widely used by statisticians for building credit scorecards, or propensity-to-buy models, or neural networks algorithms such as backpropagation, radial basis functions, or ridge polynomial networks. Others include decision trees or ensemble models such as boosted decision trees or random forests. For more complex classification problems with more than two classes, you can use multimodal techniques that predict multiple classes. Classification problems generally use supervised learning algorithms that use label data for training. Azure Machine Learning offers several algorithms for classification including logistic regression, decision trees, boosted decision trees, multimodal neural networks, etc. See Chapter 6 for more details.

Clustering Algorithms

Clustering uses unsupervised learning to group data into distinct classes. A major difference between clustering and classification problems is that the outcome of clustering is unknown beforehand. Before clustering we do not know the cluster to which each data point belongs. In contrast, with classification problems we have historical data that shows the class to which each data point belongs. For example, the lender knows from historical data whether a customer defaulted on their car loans or not.

A good application of clustering is customer segmentation where we group customers into distinct segments for marketing purposes. In a good segmentation model, the data within each segment is very similar. However, data across different segments is very different. For example, a marketer in the gaming segment needs to understand his or her customers better in order to create the right offers for them. Let's assume that he or she only has two variables on the customers: age and gaming intensity. Using clustering, the marketer finds that there are three distinct segments of gaming customers, as shown in Figure 1-7. Segment 1 is the intense gamers who play computer games passionately every day and are typically young. Segment 2 is the casual gamers who only play occasionally and are typically in their thirties or forties. The non-gamers rarely ever play computer games and are typically older; they make up Segment 3.

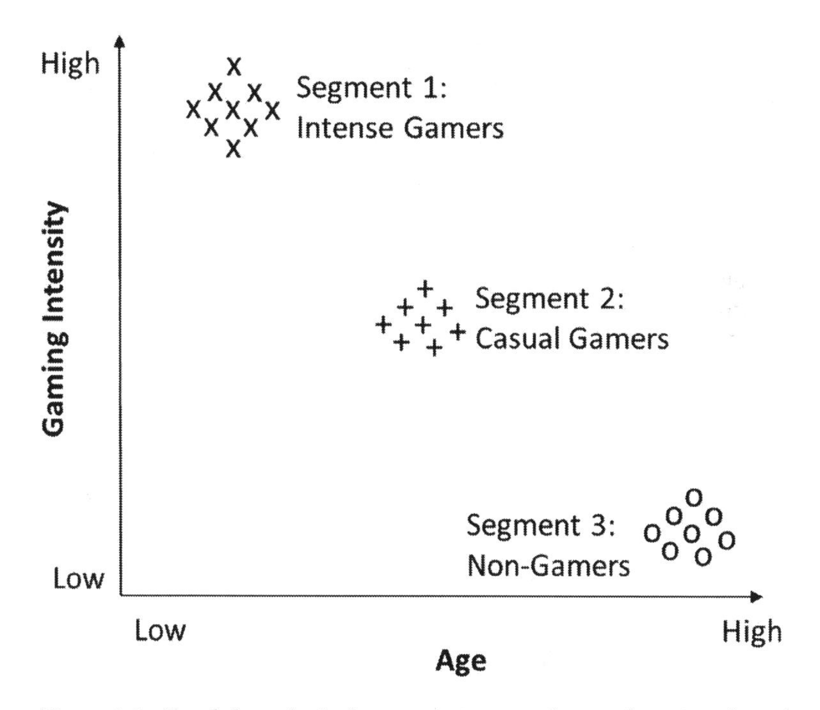

Figure 1-7. *Simple hypothetical customer segments from a clustering algorithm*

Statistics offers several tools for clustering, but the most widely used is the k-means algorithm that uses a distance metric to cluster similar data together. With this algorithm you decide a priori how many clusters you want; this is the constant K. If you set K = 3, the algorithm produces three clusters. Refer to Haralambos Marmanis and Dmitry Babenko's book for more details on the k-means algorithm. Machine learning also offers more sophisticated algorithms such as self-organizing maps (also known as Kohonen networks) developed by Teuvo Kohonen, or adaptive resonance theory (ART) networks developed by Stephen Grossberg and Gail Carpenter. Clustering algorithms typically use unsupervised learning since the outcome is not known during training.

▒ **Note** You can read more about clustering algorithms in the following books and paper:

Haralambos Marmanis and Dmitry Babenko, *Algorithms of the Intelligent Web* (Stamford, CT: Manning Publications Co., January 2011).

T. Kohonen, *Self-Organizing Maps. Third, extended edition* (Springer, 2001).

"Art2-A: an adaptive resonance algorithm for rapid category learning and recognition", Carpenter, G., Grossberg, S., and Rosen, D. Neural Networks, 4:493-504. 1991a.

Regression Algorithms

Regression techniques are used to predict response variables with numerical outcomes. For example, a wireless carrier can use regression techniques to predict call volumes at their customer service centers. With this information they can allocate the right number of call center staff to meet demand. The input variables for regression models may be numeric or categorical. However, what is common with these algorithms is that the output (or response variable) is typically numeric. Some of the most commonly used regression techniques include linear regression, decision trees, neural networks, and boosted decision tree regression.

Linear regression is one of the oldest prediction techniques in statistics and its goal is to predict a given outcome from a set of observed variables. A simple linear regression model is a linear function. If there is only one input variable, the linear regression model is the best line that fits the data. For two or more input variables, the regression model is the best hyperplane that fits the underlying data.

Artificial neural networks are a set of algorithms that mimic the functioning of the brain. They learn by example and can be trained to make predictions from a dataset even when the function that maps the response to independent variables is unknown. There are many different neural network algorithms, including backpropagation networks, Hopfield networks, Kohonen networks (also known as self-organizing maps), and adaptive resonance theory (or ART) networks. However, the most common is backpropagation, also known as multilayered perceptron. Neural networks are used for regression or classification.

Decision tree algorithms are hierarchical techniques that work by splitting the dataset iteratively based on certain statistical criteria. The goal of decision trees is to maximize the variance across different nodes in the tree, and minimize the variance within each node. Some of the most commonly used decision tree algorithms include Iterative Dichotomizer 3 (ID3), C4.5 and C5.0 (successors of ID3), Automatic Interaction Detection (AID), Chi-Squared Automatic Interaction Detection (CHAID), and Classification and Regression Tree (CART). While very useful, the ID3, C4.5, C5.0, and CHAID algorithms are classification algorithms and are not useful for regression. The CART algorithm, on the other hand, can be used for either classification or regression.

Simulation

Simulation is widely used across many industries to model and optimize processes in the real world. Engineers have long used mathematical techniques like finite elements or finite volumes to simulate the aerodynamics of aircraft wings or cars. Simulation saves engineering firms millions of dollars in R&D costs since they no longer have to do all their testing with real physical models. In addition, simulation offers the opportunity to test many more scenarios by simply adjusting variables in their computer models.

In business, simulation is used to model processes like optimizing wait times in call centers or optimizing routes for trucking companies or airlines. Through simulation, business analysts can model a vast set of hypotheses to optimize for profit or other business goals.

Statistics offers many powerful techniques for simulation and optimization. One method, the Markov chain analysis, can be used to simulate state changes in a dynamic system. For instance, it can be used to model how customers will flow through a call center: how long will a customer wait before dropping off, or what are their chances of staying on after engaging the interactive voice response (IVR) system? Linear programming is used to optimize trucking or airline routes, while Monte Carlo simulation is used to find the best conditions to optimize for given business outcome such as profit.

Content Analysis

Content analysis is used to mine content such as text files, images, and videos for insights. Text mining uses statistical and linguistic analysis to understand the meaning of text. Simple keyword searching is too primitive for most practical applications. For example, to understand the sentiment of Twitter feed data with a simple keyword search is a manual and laborious process because you have to store keywords for positive, neutral, and negative sentiments. Then, as you scan the Twitter data, you score each Twitter feed based on the specific keywords detected. This approach, though useful in narrow cases, is cumbersome and fairly primitive. The process can be automated with text mining and natural language processing (NLP), which mines the text and tries to infer the meaning of words based on context instead of simple keyword search.

Machine learning also offers several tools for analyzing images and videos through pattern recognition. Through pattern recognition, we can identify known targets with face recognition algorithms. Neural network algorithms such as multilayer perceptron and ART networks can be used to detect and track known targets in video streams, or to aid analysis of X-ray images.

Recommendation Engines

Recommendation engines have been used extensively by online retailers like Amazon to recommend products based on users' preferences. There are three broad approaches to recommendation engines. Collaboration filtering (CF) makes recommendations based on similarities between users or items. With item-based collaborative filtering, we analyze item data to find which items are similar. With collaborative filtering, that data is specifically the interactions of users with the movies, such as ratings or viewing, as opposed to characteristics of the movies such as genre, director, and actors. So whenever a customer buys a movie from this set we recommend others based on similarity.

The second class of recommendation engines makes recommendations by analyzing the content selected by each user. In this case, text mining or natural language processing techniques are used to analyze content such as document files. Similar content types are grouped together, and this forms the basis of recommendations to new users. More information on collaborative filtering and content-based approaches are available in Haralambos Marmanis and Dmitry Babenko's book.

The third approach to recommendation engines uses machine learning algorithms to determine product affinity. This approach is also known as market basket analysis. Algorithms such as Naïve Bayes, the Microsoft Association Rules, or the Arules package in R are used to mine sales data to determine which products sell together.

Cutting Edge of Data Science

Let's conclude this chapter with a quick overview of ensemble models that are at the cutting edge of data science.

The Rise of Ensemble Models

Ensemble models are a set of classifiers from machine learning that use a panel of algorithms instead of a single one to solve classification problems. They mimic our human tendency to improve the accuracy of decisions by consulting knowledgeable friends or experts. When faced with important decisions such as a medical diagnosis, we tend to seek a second opinion from other doctors to improve our confidence. In the same way, ensemble models use a set of algorithms as a panel of experts to improve the accuracy and reduce the variance of classification problems.

The machine learning community has worked on ensemble models for decades. In fact, seminal papers were published as early as 1979 by Dasarathy and Sheela. However, since the mid-1990s, this area has seen rapid progress with several important contributions resulting in very successful real-world applications.

Real-World Applications of Ensemble Models

In the last few years, ensemble models have been found in great real-world applications including face recognition in cameras, bioinformatics, Netflix movie recommendations, and Microsoft's Xbox Kinect. Let's examine two of these applications.

First, ensemble models were very instrumental to the success of the Netflix Prize competition. In 2006, Netflix ran an open contest with a $1 million prize for the best collaborative filtering algorithm that improved their existing solution by 10%. In September 2009, the $1 million prize was awarded to BellKor's Pragmatic Chaos, a team of scientists from AT&T Labs joining forces with two lesser known teams. At the start of the contest, most teams used single classifier algorithms: although they outperformed the Netflix model by 6–8%, performance quickly plateaued until teams started applying ensemble models. Leading contestants soon realized that they could improve their models by combining their algorithms with those of the apparently weaker teams. In the end, most of the top teams, including the winners, used ensemble models to significantly outperform Netflix's recommendation engine. For example, the second-place team, aptly named The Ensemble, used more than 900 individual models in their ensemble.

Microsoft's Xbox Kinect sensor also uses ensemble modeling. Random Forests, a form of ensemble model, is used effectively to track skeletal movements when users play games with the Xbox Kinect sensor.

Despite success in real-world applications, a key limitation of ensemble models is that they are black boxes in that their decisions are hard to explain. As a result, they are not suitable for applications where decisions have to be explained. Credit scorecards are a good example because lenders need to explain the credit score they assign to each consumer. In some markets, such explanations are a legal requirement and hence ensemble models would be unsuitable despite their predictive power.

Building an Ensemble Model

There are three key steps to building an ensemble model: a) selecting data, b) training classifiers, and c) combining classifiers.

The first step to build an ensemble model is data selection for the classifier models. When sampling the data, a key goal is to maximize diversity of the models, since this improves the accuracy of the solution. In general, the more diverse your models, the better the performance of your final classifier, and the smaller the variance of its predictions.

Step 2 of the process entails training several individual classifiers. But how do you assign the classifiers? Of the many available strategies, the two most popular are bagging and boosting. The bagging algorithm uses different subsets of the data to train each model. The Random Forest algorithm uses this bagging approach. In contrast, the boosting algorithm improves performance by making misclassified examples in the training set more important during training. So during training, each additional model focuses on the misclassified data. The boosted decision tree algorithm uses the boosting strategy.

Finally, once you train all the classifiers, the final step is to combine their results to make a final prediction. There are several approaches to combining the outcomes, ranging from a simple majority to a weighted majority voting.

Ensemble models are a really exciting part of machine learning, and they offer the potential for breakthroughs in classification problems.

Summary

This chapter introduced data science, defining what it is, why it matters, and why it matters now. We outlined the key academic disciplines of data science, including statistics, mathematics, operations research, signal processing, linguistics, database and storage, programming, and machine learning. We covered the key reasons behind the heightened interest in data science: increasing data volumes, data as a competitive asset, growing awareness of data mining, and hardware economics.

A simple five-step data science process was introduced with guidelines on how to apply it correctly. We also introduced some of the most commonly used techniques and algorithms in data science. Finally, we introduced ensemble models, which is one of the key technologies on the cutting edge of data science.

Bibliography

1. Alexander Linden, 2014. “Key trends and emerging technologies in advanced analytics.” Gartner BI Summit 2014, Las Vegas, USA.
2. “Are you ready for the era of Big Data?”, McKinsey Global Institute - Brad Brown, Michael Chui, and James Manyika, October 2011.
3. “Information Management in the 21st Century”, Gartner - Regina Casonato, Anne Lapkin, Mark A. Beyer, Yvonne Genovese, Ted Friedman, September 2011.
4. John C. McCallum, `www.jcmit.com/mem2012.htm`.
5. Marmanis, Haralambos and Dmitry Babenko, *Algorithms of the Intelligent Web* (Stamford, CT: Manning Publications Co., January 2011).
6. Kohonen, T., *Self-Organizing Maps. Third, extended edition* (Springer, 2001).
7. “Art2-A: an adaptive resonance algorithm for rapid category learning and recognition”, Carpenter, G., Grossberg, S., and Rosen, D. Neural Networks, 4:493-504. 1991a.
8. MacLennan, Jamie, ZhaoHui Tang, and Bogdan Crivat, *Data Mining with Microsoft SQL Server 2008* (Indianapolis, Indiana: Wiley Publishing Inc., 2009).

CHAPTER 2

Introducing Microsoft Azure Machine Learning

Azure Machine Learning, where data science, predictive analytics, cloud computing, and your data meet!

Azure Machine Learning empowers data scientists and developers to transform data into insights using predictive analytics. By making it easier for developers to use the predictive models in end-to-end solutions, Azure Machine Learning enables actionable insights to be gleaned and operationalized easily. Azure Machine Learning is a critical part of the new Cortana Analytics suite that empowers you to transform raw data into actionable insights. Chapter 14 provides more details on Cortana Analytics.

Using Machine Learning Studio, data scientists and developers can quickly build, test, and develop predictive models using state-of-the art machine learning algorithms. This chapter will provide a gentle introduction to Azure Machine Learning. You will learn the key components of an experiment and all about the Gallery of machine learning models in the product. Through a step-by-step process, you will also learn how to create models and deploy them as web services.

Azure Machine Learning has many different modules to help you build and deploy your machine learning models in production. The key steps to building and deploying your model consist of the following: 1) importing raw data; 2) data preprocessing; 3) feature engineering and data labeling (for supervised learning such as classification); 4) training, scoring, and evaluating the model; 5) model comparison and selection; 6) saving the trained model; 7) creating a predictive experiment; and 8) publishing the web service in Azure Machine Learning. Let's get started.

Hello, Machine Learning Studio!

Azure Machine Learning Studio provides an interactive visual workspace that enables you to easily build, test, and deploy predictive analytic models.

In Machine Learning Studio, you construct a predictive model by dragging and dropping datasets and analysis modules onto the design surface. You can iteratively build predictive analytic models using experiments in Azure Machine Learning Studio. Each experiment is a complete workflow with all the components required to build, test, and evaluate a predictive model. In an experiment, machine learning modules are connected together with lines that show the flow of data and parameters through the workflow. Once you design an experiment, you can use Machine Learning Studio to execute it.

Machine Learning Studio allows you to iterate rapidly by building and testing several models in minutes. When building an experiment, it is common to iterate on the design of the predictive model, edit the parameters or modules, and run the experiment several times. Often, you will save multiple copies of the experiment (using different parameters). When you first open Machine Learning Studio, you will notice it is organized as follows:

- **Experiments**: Experiments that have been created, run, and saved as drafts. These include a set of sample experiments that ship with the service to help jumpstart your projects.
- **Web Services**: A list of experiments that you have published as web services. This list will be empty until you publish your first experiment.
- **Datasets**: A list of sample datasets that ship with the product. You can use these datasets to learn about Azure Machine Learning.
- **Trained Models**: A list of any trained models that you saved from your experiments. When you first sign up, this list will be empty.
- **Settings**: A collection of settings that you can use to configure your account and resources. You can use this option to invite other users to share your workspace in Azure Machine Learning.

To develop a predictive model, you need to be able to work with data from different data sources. In addition, the data needs to be transformed and analyzed before it can be used as input for training the predictive model. Various data manipulation and statistical functions are used for preprocessing the data and identifying the parts of the data that are useful. As you develop a model, you go through an iterative process where you use various techniques to understand the data, the key features in the data, and the parameters that are used to tune the machine learning algorithms. You continuously iterate on this until you get to point where you have a trained and effective model that can be used.

Components of an Experiment

An experiment is made of the key components necessary to build, test, and evaluate a predictive model. In Azure Machine Learning, an experiment is built with a set of modules that represent datasets or different types of algorithms.

You can upload your datasets from many sources such as your local filesystem, SQL Azure, Azure Blob storage, etc. into your workspace in Azure Machine Learning. Once uploaded, you can use the dataset to create a predictive model. Machine Learning Studio also provides several sample datasets to help you jumpstart the creation of your first few experiments. As you explore Machine Learning Studio, you can upload additional datasets.

In Azure Machine Learning a module can represent either a dataset or an algorithm that you will use to build your predictive model. Machine Learning Studio provides a large set of modules to support the end-to-end data science workflow, from reading data from different data sources and preprocessing the data to building, training, scoring, and validating a predictive model. Here is an inexhaustive list of the most common modules:

- **Convert to ARFF**: Converts a .NET serialized dataset to ARFF format.
- **Convert to CSV**: Converts a .NET serialized dataset to CSV format.
- **Reader**: This module is used to read data from several sources including the Web, Azure SQL Database, Azure Blob storage, or Hive tables.
- **Writer**: This module is used to write data Azure SQL Database, Azure Blob storage, or Hadoop Distributed File system (HDFS).
- **Moving Average Filter**: This creates a moving average of a given dataset.
- **Join**: This module joins two datasets based on keys specified by the user. It does inner joins, left outer joins, full outer joins, and left semi-joins of the two datasets.
- **Split**: This module splits a dataset into two parts. It is typically used to split a dataset into separate training and test datasets.
- **Filter-Based Feature Selection**: This module is used to find the most important variables for modeling. It uses seven different techniques (Pearson Correlation, Mutual Information, Kendall Correlation, Spearman Correlation, Chi-Squared, Fisher Score, and Count-based) to rank the most important variables from raw data.
- **Elementary Statistics**: Calculates elementary statistics such as the mean, standard deviation, etc., of a given dataset.
- **Linear Regression**: Can be used to create a predictive model with a linear regression algorithm.
- **Train Model**: This module trains a selected classification or regression algorithm with a given training dataset.
- **Sweep Parameters**: For a given learning algorithm, along with training and validation datasets, this module finds parameters that result in the best trained model.
- **Evaluate Model**: This module is used to evaluate the performance of a trained classification or regression model.

- **Cross Validate Model**: This module is used to perform cross-validation to avoid over-fitting. By default this module uses 10-fold cross-validation.
- **Score Model**: Scores a trained classification or regression model.

All available modules are organized under the menus shown in Figure 2-1. Each module provides a set of parameters that you can use to fine-tune the behavior of the algorithm used by the module. When a module is selected, you will see the parameters for the module displayed on the right pane of the canvas.

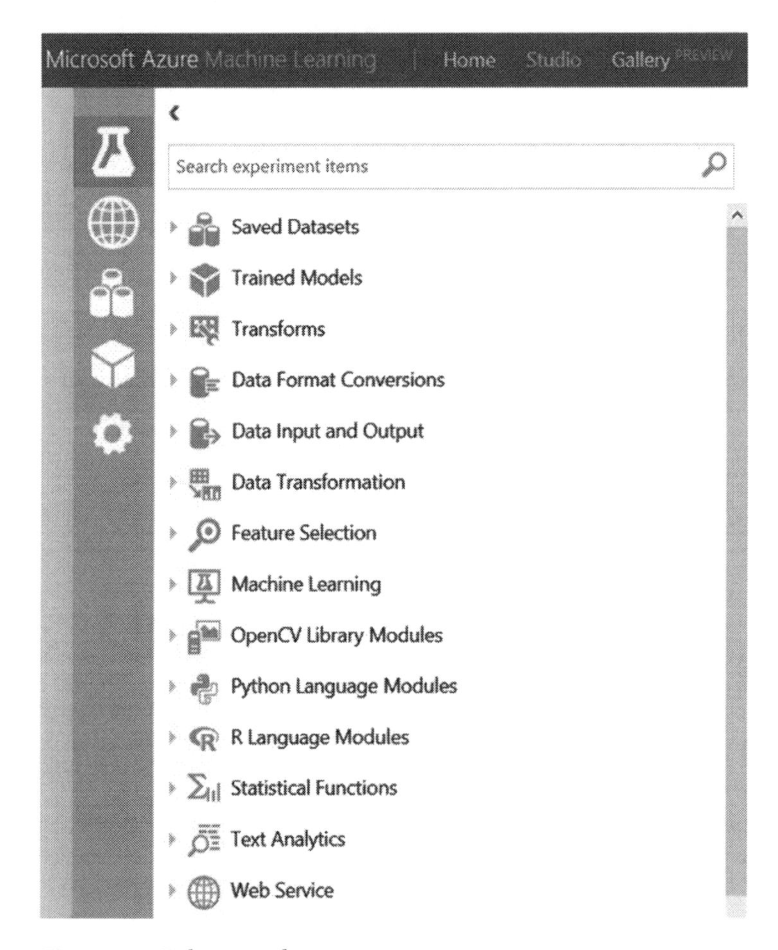

Figure 2-1. *Palette search*

Introducing the Gallery

Azure Machine Learning provides a gallery of sample machine learning models you can use to quickly get started. Each of these models contains an experiment provided by Microsoft, its partners, or individual data scientists. In time you too can publish your own models in this gallery if you choose to share your work with the data science community. Figure 2-2 shows a screenshot of the Gallery. You can also find the Gallery at `https://gallery.azureml.net/`.

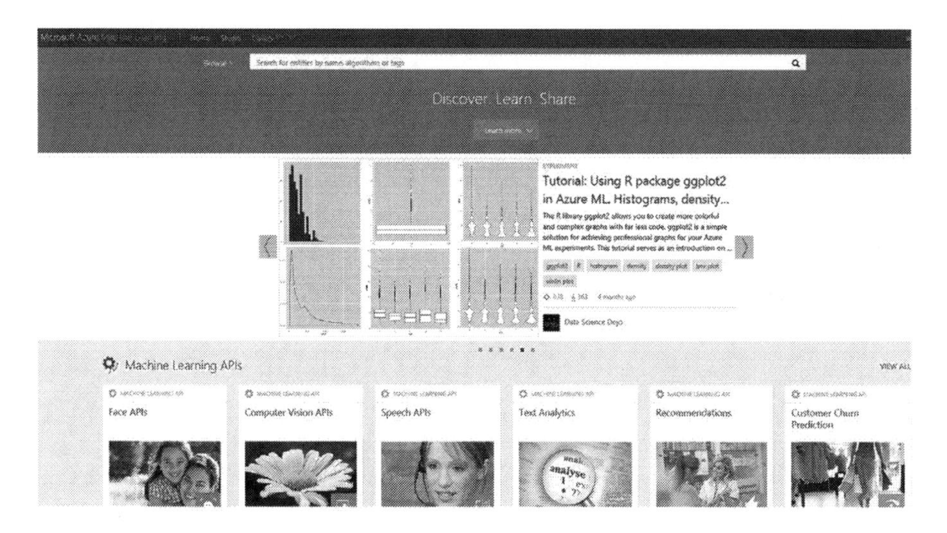

Figure 2-2. *The Gallery in Azure Machine Learning*

The Gallery is a great way to get started quickly since it offers sample solutions to many common problems such as fraud detection, face recognition, recommendations, churn analysis, or predictive maintenance. These samples serve as templates you can use to kickstart your own projects instead of starting from scratch. For example, if you are building a new model for churn analysis, the templates in the gallery offer sample experiments that show the key steps and the right modules you can use in your own experiment. The following four experiments in this book are published in the Gallery:

- Customer propensity model in Chapter 7
- Customer churn model discussed in Chapter 9
- Customer segmentation model in Chapter 10
- Predictive maintenance model covered in Chapter 11

Five Easy Steps to Creating a Training Experiment

In this section, you will learn how to use Azure Machine Learning Studio to develop a simple predictive analytics model. To design an experiment, you assemble a set of components that are used to create, train, test, and evaluate the model. In addition, you might leverage additional modules to preprocess the data, perform feature selection and/or reduction, split the data into training and test sets, and evaluate or cross-validate the model. The following five basic steps can be used as a guide for creating an experiment.

Create a Model

Step 1: Get data

Step 2: Preprocess data

Step 3: Define features

Train the Model

Step 4: Choose and apply a learning algorithm

Test the Model

Step 5: Predict over new data

Step 1: Getting the Data

Azure Machine Learning Studio provides a number of sample datasets. In addition, you can import data from many different sources. In this example, you will use the included sample dataset called **Automobile price data (Raw)**, which contains automobile price data.

1. To start a new experiment, click **+NEW** at the bottom of the Machine Learning Studio window and select **EXPERIMENT**. This gives you the option to start with a blank experiment, follow a tutorial, or choose from the list of sample experiments from the Gallery you saw in the previous section.

2. Select **Blank Experiment** and rename the experiment to `Chapter 02 - Hello ML`.

3. To the left of the Machine Learning Studio is a list of experiment items (see Figure 2-1). Click **Saved Datasets**, and type "automobile" in the search box. Find **Automobile price data (Raw)**.

4. Drag the dataset into the experiment. You can also double-click the dataset to include it in the experiment (see Figure 2-3).

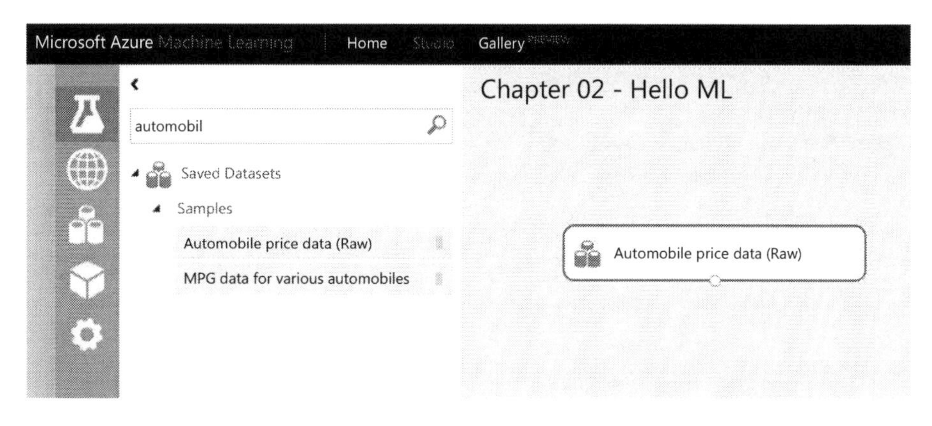

Figure 2-3. *Using a dataset*

By right-clicking the output port of the dataset, you can select **Visualize**, which will allow you to explore the data and understand the key statistics of each of the columns (see Figure 2-4).

Chapter 02 – Hello ML › Automobile price data (Raw) › dataset

rows 205 columns 26

symboling	normalized-losses	make	fuel-type	aspiration	num-of-doors	body-st
3		alfa-romero	gas	std	two	convert
3		alfa-romero	gas	std	two	convert
1		alfa-romero	gas	std	two	hatchb
2	164	audi	gas	std	four	sedan
2	164	audi	gas	std	four	sedan
2		audi	gas	std	two	sedan
1	158	audi	gas	std	four	sedan
1		audi	gas	std	four	wagon
1	158	audi	gas	turbo	four	sedan
0		audi	gas	turbo	two	hatchb

Statistics

Mean	122
Median	115
Min	65
Max	256
Standard Deviation	35.4422
Unique Values	51
Missing Values	41
Feature Type	Numeric Feature

Visualizations

normalized-losses
Histogram
compare to None

Figure 2-4. *Dataset visualization*

The goal of this exercise is to build a predictive model that uses features of the **Automobile price data (Raw)** dataset to predict the price for new automobiles. Each row represents an automobile. Each column is a feature of that automobile. For example, make is the make of the car, and fuel-type indicates if the car uses gas or diesel.

Close the visualization window by clicking the **x** in the upper-right corner.

Step 2: Preprocessing the Data

Before you start designing the experiment, it is important to preprocess the dataset. In most cases, the raw data needs to be preprocessed before it can be used as input to train a predictive analytic model.

From the earlier exploration, you may have noticed that there are missing values in the data. As a precursor to analyzing the data, these missing values need to be cleaned. For this experiment, you will substitute the missing values with a designated value. In addition, the `normalized-losses` column will be removed as this column contains too many missing values.

■ **Tip** Cleaning the missing values from input data is a prerequisite for using most of the modules.

---

1. To remove the `normalized-losses` column, drag the **Project Columns** module, and connect it to the output port of the **Automobile price data (Raw)** dataset. This module allows you to select which columns of data you want to include or exclude in the model.

2. Select the **Project Columns** module and click **Launch column selector** in the properties pane (the right pane).

    a. Make sure **All columns** is selected in the filter drop-down called **Begin With**. This directs Project Columns to pass all columns through (except for the ones you are about to exclude).

    b. In the next row, select **Exclude** and **column names**, and then click inside the text box. A list of columns is displayed; select `normalized-losses` and it will be added to the text box. This is shown in Figure 2-5.

    c. Click the check mark OK button to close the column selector.

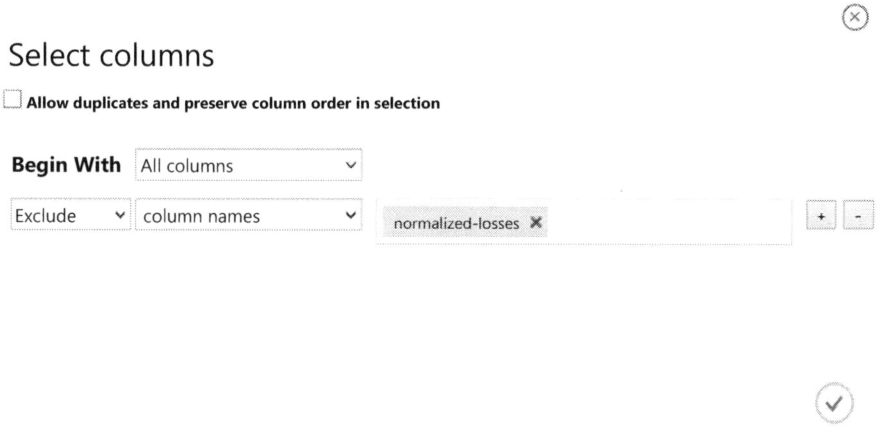

Select columns

Allow duplicates and preserve column order in selection

Begin With All columns

Exclude column names normalized-losses

Figure 2-5. *Select columns*

All columns will pass through, except for the column `normalized-losses`. You can see this in the properties pane for **Project Columns**. This is illustrated in Figure 2-6.

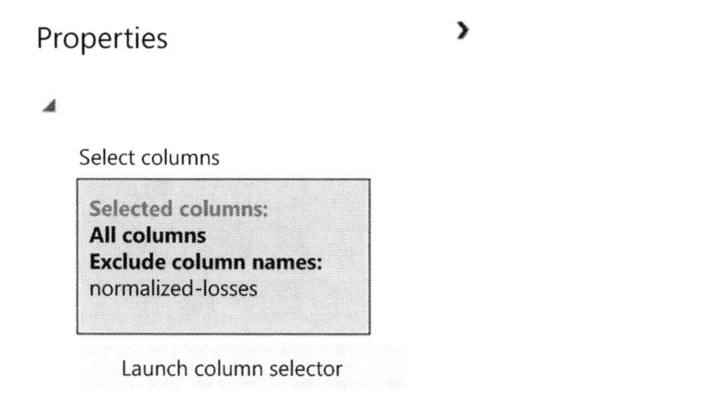

Figure 2-6. *Project Columns properties*

■ **Tip** As you design the experiment, you can add a comment to the module by double-clicking the module and entering text. This enables others to understand the purpose of each module in the experiment and can help you document your experiment design.

---

3. Drag the **Clean Missing Data** module to the experiment canvas and connect it to the **Project Columns** module. You will use the default properties; this replaces the missing value with a 0. See Figure 2-7 for details.

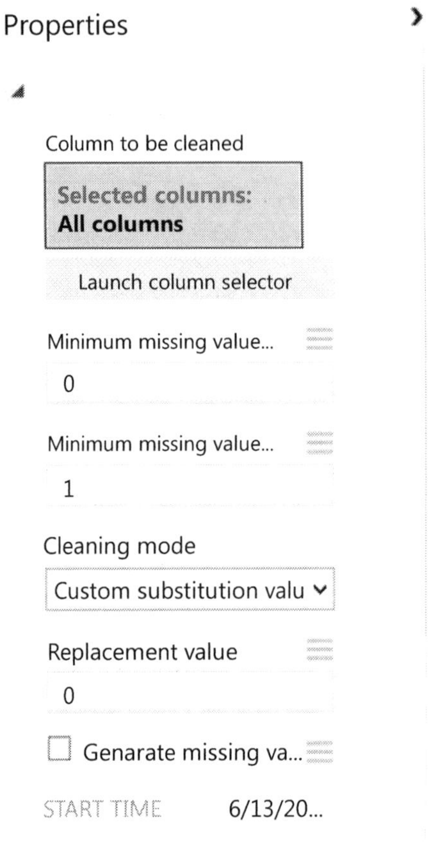

Properties

Column to be cleaned

Selected columns:
All columns

Launch column selector

Minimum missing value...

0

Minimum missing value...

1

Cleaning mode

Custom substitution valu

Replacement value

0

Genarate missing va...

START TIME 6/13/20...

Figure 2-7. *Missing Values Scrubber properties*

4. Now click **RUN**.
5. When the experiment completes successfully, each of the modules will have a green check mark indicating its successful completion (see Figure 2-8).

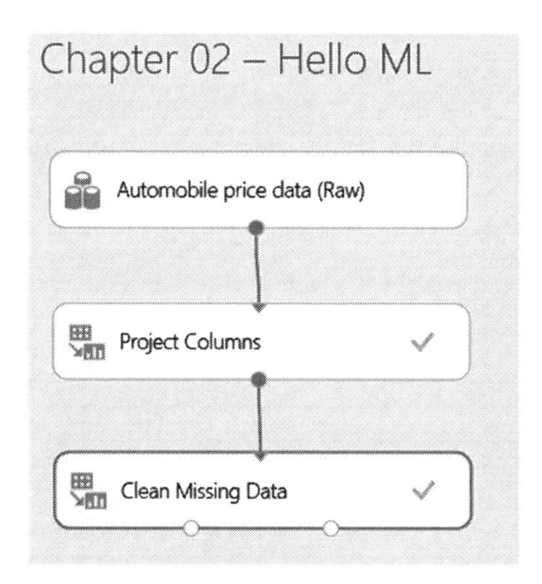

Figure 2-8. *First experiment run*

At this point, you have preprocessed the dataset by cleaning and transforming the data. To view the cleaned dataset, double-click the output port of the **Clean Missing Data** module and select **Visualize**. Notice that the `normalized-losses` column is no longer included, and there are no missing values.

Step 3: Defining the Features

In machine learning, features are individual measurable properties created from the raw data to help the algorithms to learn the task at hand. Understanding the role played by each feature is super important. For example, some features are better at predicting the target than others. In addition, some features can have a strong correlation with other features (e.g. city-mpg vs. highway-mpg). Adding highly correlated features as inputs might not be useful, since they contain similar information.

For this exercise, you will build a predictive model that uses a subset of the features of the **Automobile price data (Raw)** dataset to predict the price for new automobiles. Each row represents an automobile. Each column is a feature of that automobile. It is important to identify a good set of features that can be used to create the predictive model. Often, this requires experimentation and knowledge about the problem domain. For illustration purpose, you will use the **Project Columns** module to select the following features: `make, body-style, wheel-base, engine-size, horsepower, peak-rpm, highway-mpg, and price.`

1. Drag a second **Project Columns** module to the experiment canvas. Connect it to the **Clean Missing Data** module.
2. Click **Launch column selector** in the properties pane.
3. In the column selector, select **No columns** for **Begin With**, then select **Include** and **column names** in the filter row. Enter the following column names: `make, body-style, wheel-base, engine-size, horsepower, peak-rpm, highway-mpg,` and `price`. This directs the module to pass through only these columns.
4. Click **OK**.

■ **Tip** As you build the experiment, you will run it. By running the experiment, you enable the column definitions of the data to be used in the **Clean Missing Data** module.

When you connect **Project Columns** to **Clean Missing Data**, the **Project Columns** module becomes aware of the column definitions in your data. When you click the column names box, a list of columns is displayed and you can then select the columns, one at a time, that you wish to add to the list.

Figure 2-9 shows the list of selected columns in the **Project Columns** module. When you train the predictive model, you need to provide the target variable. This is the feature that will be predicted by the model. For this exercise, you are predicting the price of an automobile, based on several key features of an automobile (e.g. horsepower, make, etc.)

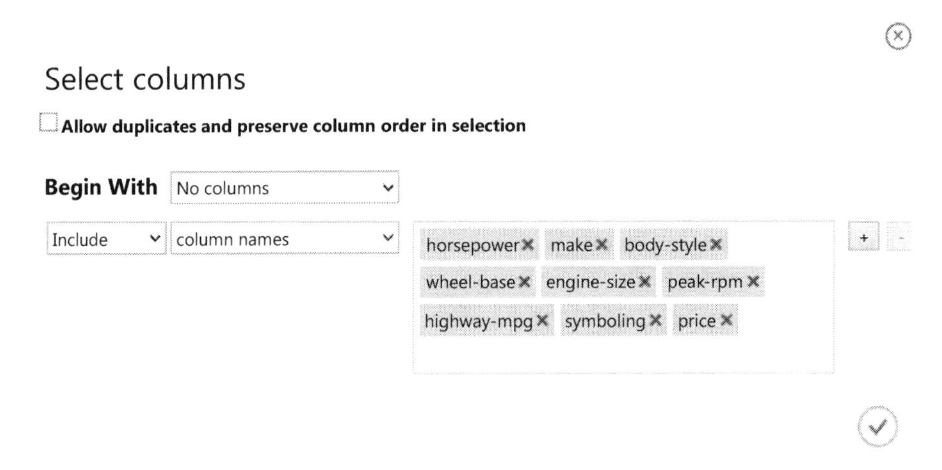

Figure 2-9. *Select columns*

Step 4: Choosing and Applying Machine Learning Algorithms

When constructing a predictive model, you first need to train the model, and then validate that the model is effective. In this experiment, you will build a regression model.

■ **Tip** Classification and regression are two common types of predictive models. In classification, the goal is to predict if a given data row belongs to one of several classes (Will a customer churn or not? Is this credit transaction fraudulent?). With regression, the goal is to predict a continuous outcome (such as the price of an automobile or tomorrow's temperature).

In this experiment, you will train a regression model and use it to predict the price of an automobile. Specifically, you will train a simple *linear regression* model. After the model has been trained, you will use some of the modules available in Machine Learning Studio to validate the model.

1. **Split the data into training and testing sets**: Select and drag the **Split** module to the experiment canvas and connect it to the output of the last **Project Columns** module. Set **Fraction of rows in the first output dataset** to 0.8. This way, you will use 80% of the data to train the model and hold back 20% for testing.

■ **Tip** By changing the Random seed parameter, you can produce different random samples for training and testing. This parameter controls the seeding of the pseudo-random number generator in the **Split** module.

---

2. Run the experiment. This allows the **Project Columns** and **Split** modules to pass along column definitions to the modules you will be adding next.

3. To select the learning algorithm, expand the **Machine Learning** category in the module palette to the left of the canvas and then expand **Initialize Model**. This displays several categories of modules that can be used to initialize a learning algorithm.

4. For this example experiment, select the **Linear Regression** module under the **Regression** category and drag it to the experiment canvas.

5. Find and drag the **Train Model** module to the experiment. Click **Launch column selector** and select the `price` column. This is the feature that your model is going to predict. Figure 2-10 shows this target selection.

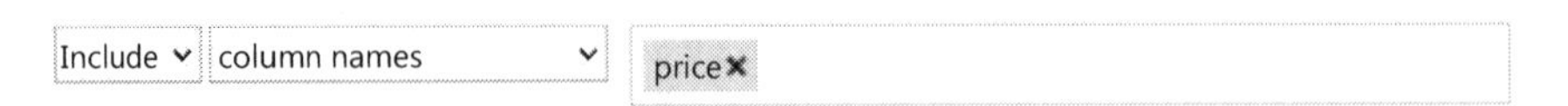

Figure 2-10. *Select the price column*

6. Connect the output of the **Linear Regression** module to the left input port of the **Train Model** module.
7. Also, connect the training data output (the left port) of the **Split** module to the right input port of the **Train Model** module.
8. Run the experiment.

The result is a trained regression model that can be used to score new samples to make predictions. Figure 2-11 shows the experiment up to this point.

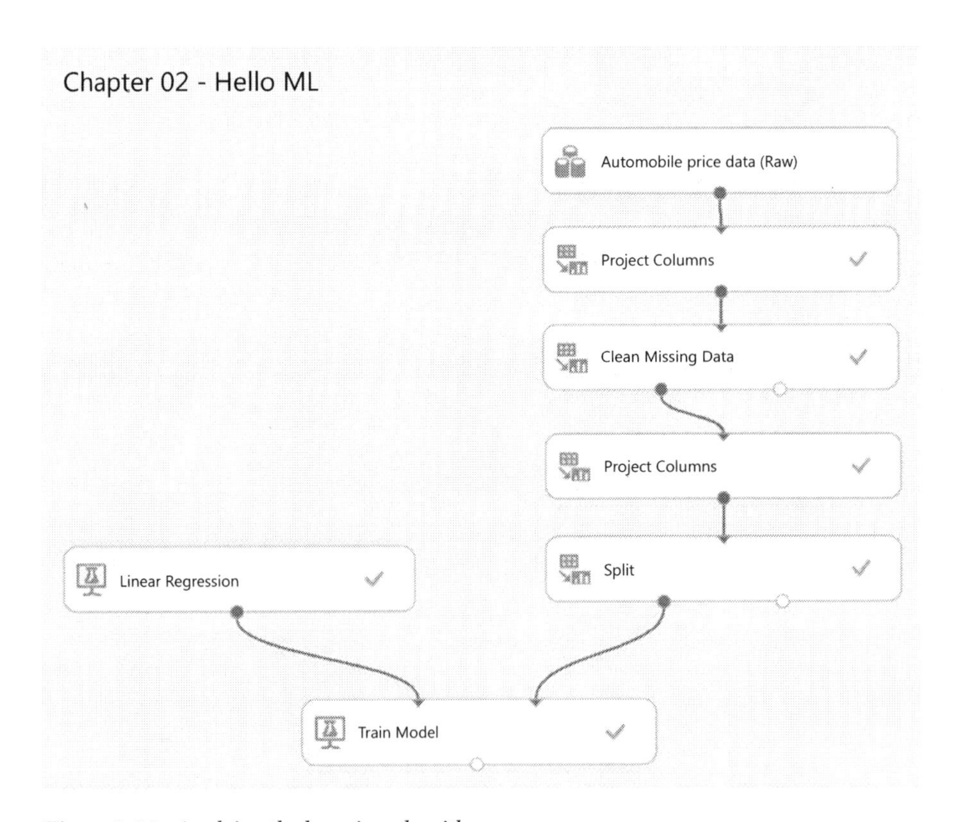

Figure 2-11. *Applying the learning algorithm*

Step 5: Predicting Over New Data

Now that you've trained the model, you can use it to score the other 20% of your data and see how well your model predicts on unseen data.

1. Find and drag the **Score Model** module to the experiment canvas and connect the left input port to the output of the **Train Model** module, and the right input port to the test data output (right port) of the **Split** module. See Figure 2-12 for details.

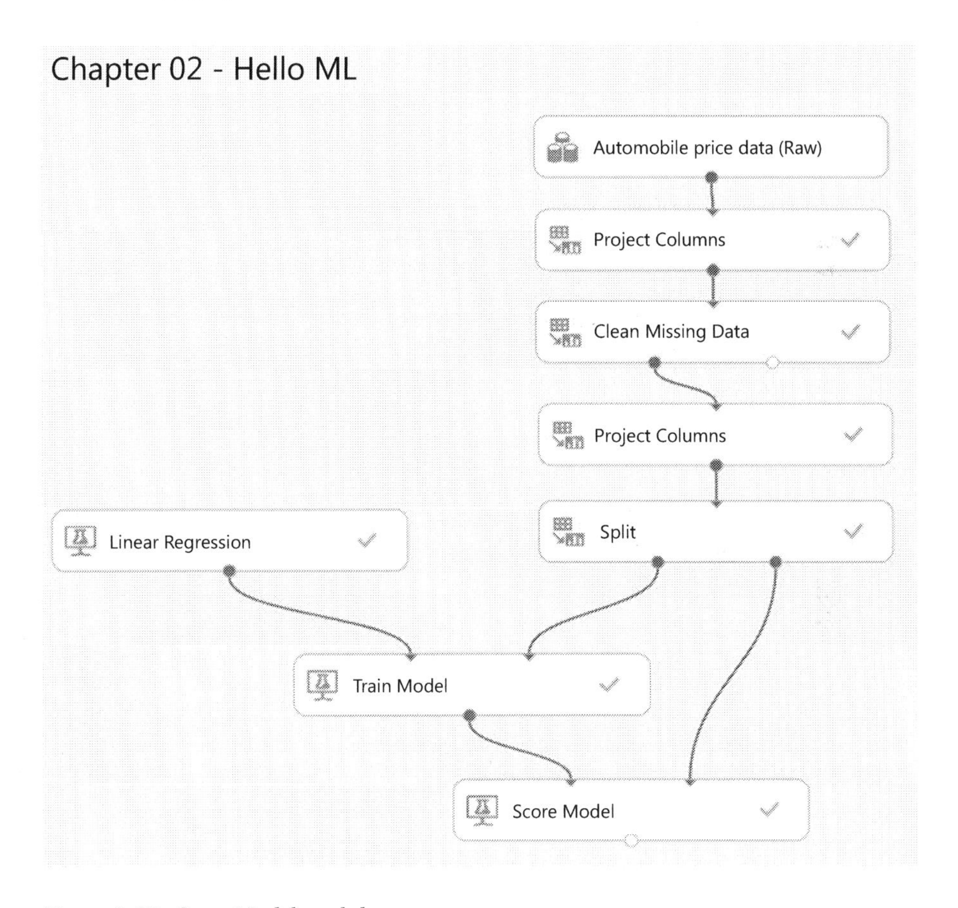

Figure 2-12. *Score Model module*

2. Run the experiment and view the output from the **Score Model** module (by right-clicking the output port and selecting **Visualize**). The output will show the predicted values for price along with the known values from the test data.

3. Finally, to test the quality of the results, select and drag the **Evaluate Model** module to the experiment canvas, and connect the left input port to the output of the **Score Model** module (there are two input ports because the **Evaluate Model** module can be used to compare two different models).

4. Run the experiment and view the output from the **Evaluate Model** module (right-click the output port and select **Visualize**). The following statistics are shown for your model:

    a. **Mean Absolute Error** (MAE): The average of absolute errors (an *error* is the difference between the predicted value and the actual value).

    b. **Root Mean Squared Error** (RMSE): The square root of the average of squared errors.

    c. **Relative Absolute Error**: The average of absolute errors relative to the absolute difference between actual values and the average of all actual values.

    d. **Relative Squared Error**: The average of squared errors relative to the squared difference between the actual values and the average of all actual values.

    e. **Coefficient of Determination**: Also known as the R squared value, this is a statistical metric indicating how well a model fits the data.

For each of the error statistics, smaller is better; a smaller value indicates that the predictions more closely match the actual values. For **Coefficient of Determination**, the closer its value is to one (1.0), the better the predictions (see Figure 2-13). If it is 1.0, this means the model explains 100% of the variability in the data, which is pretty unrealistic!

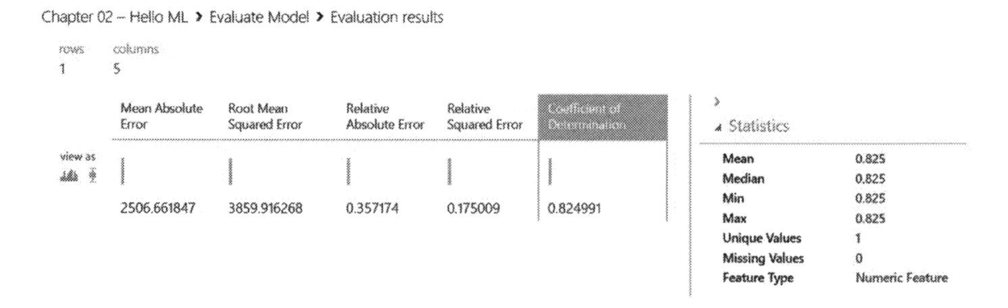

***Figure 2-13.** Evaluation results*

The final experiment should look like Figure 2-14.

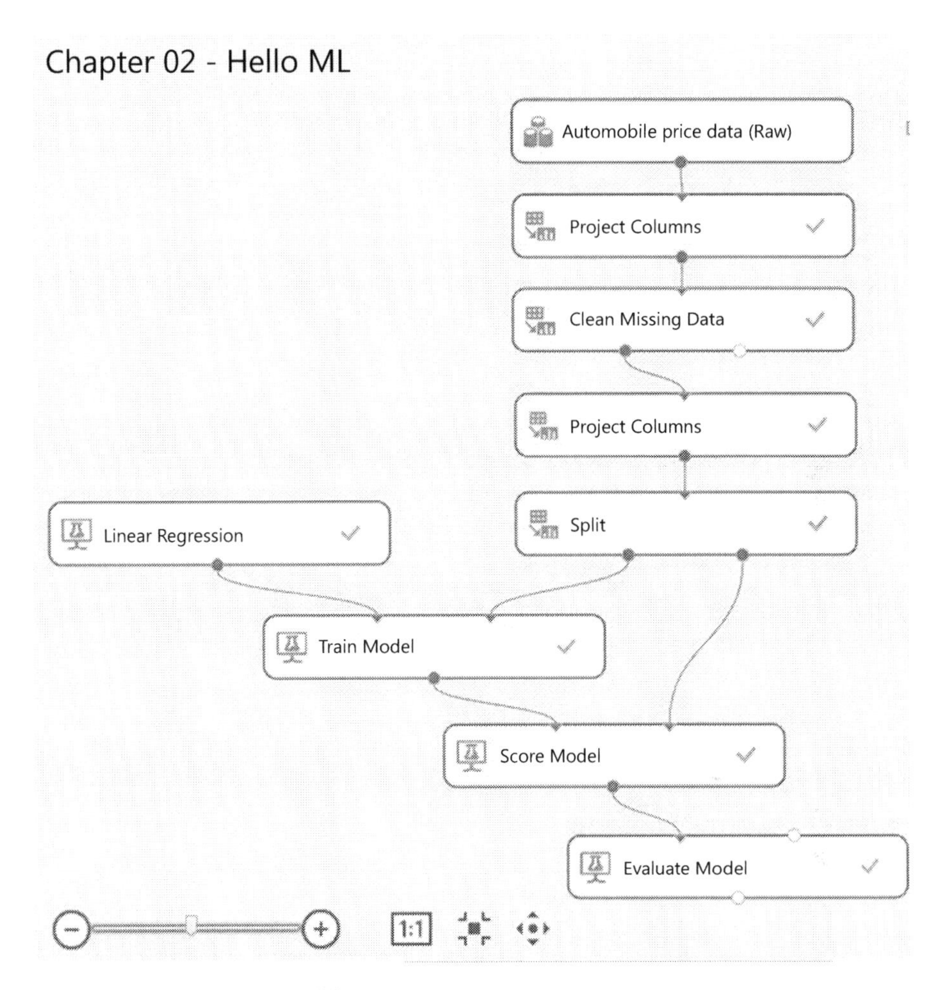

Figure 2-14. *Regression Model experiment*

Congratulations! You have created your first machine learning experiment in Machine Learning Studio. In Chapters 7-12, you will see how to apply these five steps to create predictive analytics solutions that address business challenges from different domains such as buyer propensity, churn analysis, customer segmentation, and predictive maintenance. In addition, Chapter 4 shows how to use R scripts as part of your experiments in Azure Machine Learning. In Chapter 5, you will learn how to use Python as part of your experiment in Azure Machine Learning.

Deploying Your Model in Production

Today it takes too long to deploy machine learning models in production. The process is typically inefficient and often involves rewriting the model to run on the target production platform, which is costly and requires considerable time and effort. Azure Machine Learning simplifies the deployment of machine learning models through an integrated process in the cloud. You can deploy your new predictive model in a matter of minutes instead of days or weeks. Once deployed, your model runs as a web service that can be called from different platforms including servers, laptops, tablets, or even smartphones. To deploy your model in production, follow these two steps.

1. Create a predictive experiment.
2. Publish your experiment as a web service.

Creating a Predictive Experiment

To create a predictive experiment, follow these steps in Azure Machine Learning Studio.

1. Run your experiment with the **Run** button at the bottom of Azure Machine Learning Studio.
2. Select the **Train Model** module in your experiment. This tells the tools that you plan to deploy the Linear Regression model in production. This step is only necessary if you have several training modules in your experiment
3. Next, click **Setup Web Service ➤ Predictive Web Service (Recommended)** at the bottom of Azure Machine Learning Studio. Azure Machine Learning will automatically create a predictive experiment. In the process, it deletes all the modules that are not needed. For example, the other Train Model modules plus the Split, Project, and other modules are removed. The tool replaces the **Linear Regression** module and its Train Model module with the newly trained model. It will also add a new web input and output for your experiment.
4. The **Predictive experiment** that is created is called `Chapter 02 - Hello ML [Scoring Exp.]`.
5. Your **Predictive experiment** should appear as shown Figure 2-15. You are now ready to publish your web service.

■ **Tip** You may be wondering why you left the **Automobile price data (Raw)** dataset connected to the model. The service is going to use the user's data, not the original dataset, so why leave them connected?

It's true that the service doesn't need the original automobile price data. But it does need the schema for that data, which includes information such as the number of columns and which columns are numeric. This schema information is necessary in order to interpret the user's data. You leave these components connected so that the scoring module will have the dataset schema when the service is running. The data isn't used, just the schema.

Figure 2-15. The experiment that uses the saved training model

---

■ **Tip** You can update the web service after you've published it. For example, if you want to change your model, just edit the training experiment you saved earlier, tweak the model parameters, and save the trained model (overwriting the one you saved before). After you have re-run the training experiment, you can choose **Update Predictive Experiment**. This will replace the web service, using your newly trained model.

You can also choose to create a Retraining web service, which will enable you to invoke it to retrain the model.

Publishing Your Experiment as a Web Service

At this point, your model is now ready to be deployed as a web service. To do this, follow these steps.

1. Run your experiment with the **Run** button at the bottom of Azure Machine Learning Studio.

2. Click **Deploy Web Service** button at the bottom of Azure Machine Learning Studio. The tool will deploy your predictive experiment as a web service. The result should appear as shown in Figure 2-16.

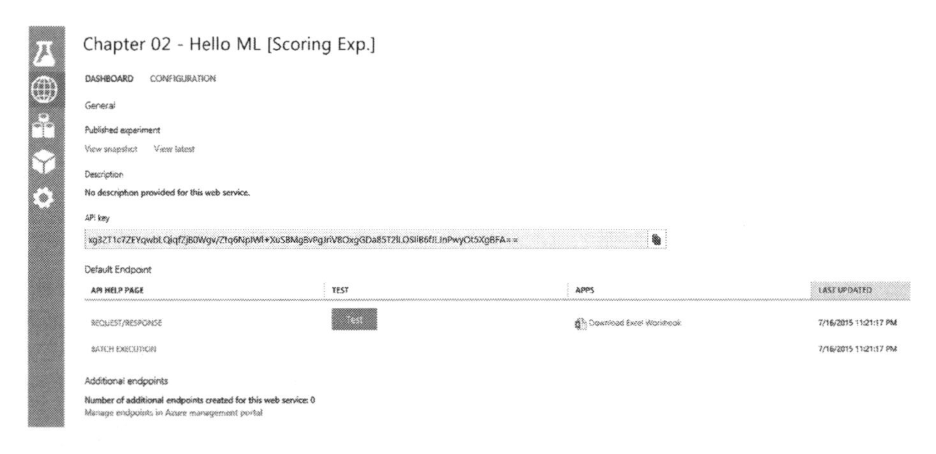

Figure 2-16. *A dialog box that promotes the machine learning model from the staging server to a live production web service*

Congratulations, you have just published your machine learning model into production. Figure 2-16 shows the API key for your model as well as the URLs you will use to call your model either interactively in request/response mode, or in batch mode. It also shows a link to a new Excel workbook you can download to your local file system. With this spreadsheet you can call your model to score data in Excel. In addition, you can also use the sample code provided to invoke your new web service in C#, Python, or R.

Accessing the Azure Machine Learning Web Service

To be useful as a web service, users need to be able to send data to the service and receive results. The web service is an Azure web service that can receive and return data in one of two ways:

- **Request/Response**: The user can send a single set of Automobile price data to the service using an HTTP protocol, and the service responds with a single result predicting the price.
- **Batch Execution**: The user can send to the service the URL of an Azure BLOB that contains one or more rows of Automobile price data. The service stores the results in another BLOB and returns the URL of that container.

On the **Dashboard** tab for the web service, there are links to information that will help a developer write code to access this service. Click the API help page link on the REQUEST/RESPONSE row and a page opens that contains sample code to use the service's request/response protocol. Similarly, the link on the BATCH EXECUTION row provides example code for making a batch request to the service.

The API help page includes samples for R, C#, and Python programming languages. For example, Listing 2-1 shows the R code that you could use to access the web service you published (the actual service URL would be displayed in your sample code). Before using this sample code you will need to replace the API key of "abc123" with the real API key shown in Figure 2-16.

Listing 2-1. R Code Used to Access the Service Programmatically

```
library("RCurl")
library("rjson")

# Accept SSL certificates issued by public Certificate Authorities
options(RCurlOptions = list(cainfo = system.file("CurlSSL", "cacert.pem",
package = "RCurl")))

h = basicTextGatherer()
hdr = basicHeaderGatherer()

req = list(

        Inputs = list(

            "input1" = list(
                "ColumnNames" = list("symboling", "normalized-losses",
"make", "fuel-type", "aspiration", "num-of-doors", "body-style",
"drive-wheels", "engine-location", "wheel-base", "length", "width",
"height", "curb-weight", "engine-type", "num-of-cylinders", "engine-size",
"fuel-system", "bore", "stroke", "compression-ratio", "horsepower",
"peak-rpm", "city-mpg", "highway-mpg", "price"),
```

```
                "Values" = list( list( "0", "0", "value", "value", "value",
"value", "value", "value", "value", "0", "0", "0", "0", "0", "value",
"value", "0", "value", "0", "0", "0", "0", "0", "0", "0", "0" ), list( "0",
"0", "value", "value", "value", "value", "value", "value", "value", "0",
"0", "0", "0", "0", "value", "value", "0", "value", "0", "0", "0", "0", "0",
"0", "0", "0" ) )
            )           ),
        GlobalParameters = fromJSON('{}')
)

body = enc2utf8(toJSON(req))
api_key = "abc123" # Replace this with the API key for the web service
authz_hdr = paste('Bearer', api_key, sep=' ')

h$reset()
curlPerform(url = "https://ussouthcentral.services.azureml.net/workspaces/
fcaf778fe92f4fefb2f104acf9980a6c/services/1dc53e8e4eb54288838b65bee7911541/
execute?api-version=2.0&details=true",
            httpheader=c('Content-Type' = "application/json",
'Authorization' = authz_hdr),
            postfields=body,
            writefunction = h$update,
            headerfunction = hdr$update,
            verbose = TRUE
            )

headers = hdr$value()
httpStatus = headers["status"]
if (httpStatus >= 400)
{
    print(paste("The request failed with status code:", httpStatus, sep="
"))

    # Print the headers - they include the requert ID and the timestamp,
which are useful for debugging the failure
    print(headers)
}

print("Result:")
result = h$value()
print(fromJSON(result))
```

Summary

In this chapter, you used Azure Machine Learning Studio to create your first experiment. You learned how to perform data preprocessing, and how to train, test, and evaluate your model in Azure Machine Learning Studio. In addition, you also saw how to deploy your predictive experiment as a web service. Once deployed, your predictive experiment runs as a web service on Azure that can be called from different applications. Sample code in C#, R and Python is provided to enable you to get started quickly.

In the remainder of this book, you will learn how to use Azure Machine Learning to create experiments that solve various business problems such as customer propensity, customer churn, recommendations, and predictive maintenance. In addition, you will also learn how to extend Azure Machine Learning with R and Python scripting. Also, Chapter 6 introduces the most commonly used statistics and machine learning algorithms in Azure Machine Learning.

CHAPTER 3

Data Preparation

Machine learning can feel magical. You provide Azure ML with training data, select an appropriate leaning algorithm, and it can learn patterns in that data. In many cases, the performance of the model that you build, if done correctly, will outperform a human expert. But, like so many problems in the world, there is a significant "garbage in, garbage out" aspect to machine learning. If the data you give it is rubbish, the learning algorithm is unlikely to be able to overcome it. Machine learning can't perform "data alchemy" and turn data lead into gold; that's why we practice good data science, and first clean and enhance the data so that the learning algorithm can do its magic. Done correctly, it's the perfect collaboration between data scientist and machine learning algorithms.

In this chapter, we will describe the following three different methods in which the input data can be processed using Azure ML to prepare for machine learning methods and improve the quality of the resulting model:

1. **Data Cleaning and Processing:** In this step, you ensure that the collected data is clean and consistent. It includes tasks such as integrating multiple datasets, handling missing data, handling inconsistent data, and converting data types.

2. **Feature Selection:** In this step, you select the key subset of original data features in an attempt to reduce the dimensionality of the training problem.

3. **Feature Engineering:** In this step, you create additional relevant features from the existing raw features in the data that increase the predictive power of the resulting model.

Cleaning and processing your training data, followed by selecting and engineering features, can increase the efficiency of the machine learning algorithm in its attempt to extract key information contained in the data. Mathematically speaking, the features used to train the model should be the minimal set of independent variables that explain the patterns in the data and then predict outcomes successfully. This can improve the power of the resulting model to more accurately predict data it has never seen.

Throughout this chapter we will illustrate the procedures involved in each step using Azure Machine Learning Studio. In each example, we will use either sample datasets available in Azure ML or public data from the UCI machine learning data repository, which you can upload to Azure ML so that you can recreate each step to gain first-hand experience.

Data Cleaning and Processing

Machine learning algorithms learn from data. It is critical that you feed them the right data for the problem you wish to solve. Even if you have good data, you need to make sure that it is in a useful scale, the right format, and even that meaningful features are included. In this section, you will learn how to prepare data for machine learning. This is a big topic and we will cover the essentials, starting with getting to know your data.

Getting to Know Your Data

There is no substitute for getting to know your data. Real world data is never clean. Raw data is often noisy, unreliable, and may be missing values. Using such data for modeling will likely produce misleading results. Visualizing histograms of the distribution of values of nominal attributes and plotting the values of numeric attributes are extremely helpful to provide you with insights into the data you are working with. A graphical display of the data also makes it easy to identify outliers, which may very well represent errors in the data file; this allows you to identify inconsistent data, and can let you know if you have missing data to deal with. With a large dataset, you may be tempted to give up; how can you possibly check it all? This is where the ability for Azure Machine Learning to create a visualization of a dataset, along with the ability to sample from an extremely large dataset, shines. Before diving straight into that, let's first consider what you might want to learn from your first examination of a dataset. Examples include

- The number of records in the dataset
- The number of attributes (or features)
- For each attribute, what are the data types (nominal, ordinal, or continuous)
- For nominal attributes, what are the values
- For continuous attribute, the statistics and distribution
- For any attributes, the number of missing values
- For any attribute, the number of distinct values
- For labeled data, check that the classes approximately balanced
- For inconsistent data records, for a given attribute, check that values are within the expected range

Let's begin by visualizing data in a dataset. For this example, you will use the Automobile price dataset, so drag it into your experiment and right-click the output pin. This brings up options, including the ability to visualize the dataset, as illustrated in Figure 3-1; select this option.

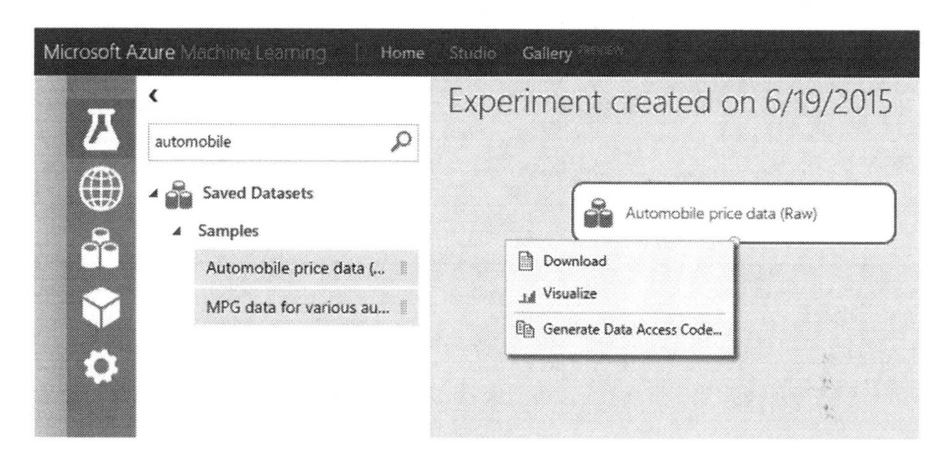

Figure 3-1. *Visualizing a dataset in Azure ML Studio*

The result are returned in an information-rich infographic (Figure 3-2). In the upper left corner are the number of rows in the dataset, along with the number of columns. Below that are the names of each column (attribute), along with a histogram of the attribute distribution and below this histogram is a list of the values with missing values are represented by blank spaces; there is an option to switch from histograms to box plots for numeric attributes. Immediately you have learned a great deal of information about your dataset.

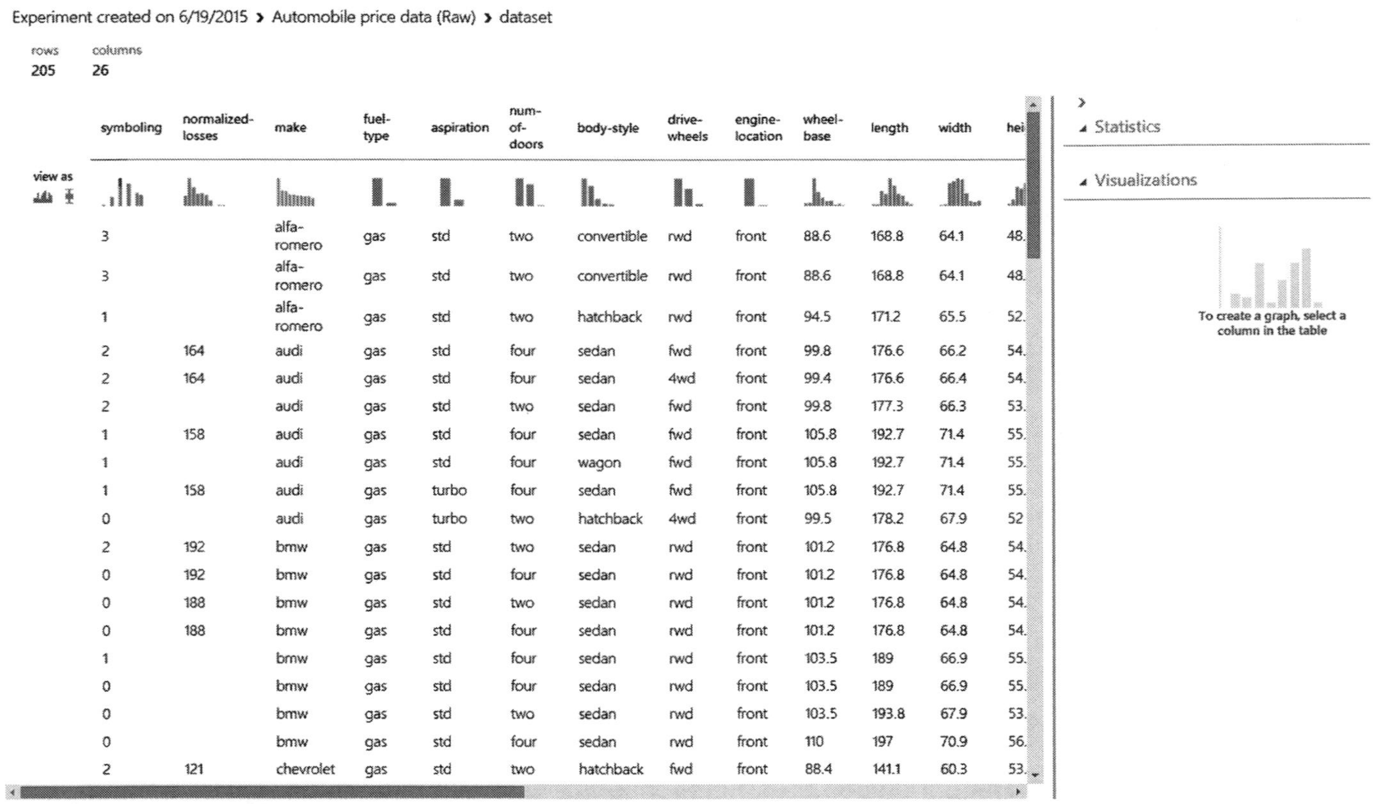

Experiment created on 6/19/2015 › Automobile price data (Raw) › dataset

rows 205 columns 26

symboling	normalized-losses	make	fuel-type	aspiration	num-of-doors	body-style	drive-wheels	engine-location	wheel-base	length	width	hei
3		alfa-romero	gas	std	two	convertible	rwd	front	88.6	168.8	64.1	48.
3		alfa-romero	gas	std	two	convertible	rwd	front	88.6	168.8	64.1	48.
1		alfa-romero	gas	std	two	hatchback	rwd	front	94.5	171.2	65.5	52.
2	164	audi	gas	std	four	sedan	fwd	front	99.8	176.6	66.2	54.
2	164	audi	gas	std	four	sedan	4wd	front	99.4	176.6	66.4	54.
2		audi	gas	std	two	sedan	fwd	front	99.8	177.3	66.3	53.
1	158	audi	gas	std	four	sedan	fwd	front	105.8	192.7	71.4	55.
1		audi	gas	std	four	wagon	fwd	front	105.8	192.7	71.4	55.
1	158	audi	gas	turbo	four	sedan	fwd	front	105.8	192.7	71.4	55.
0		audi	gas	turbo	two	hatchback	4wd	front	99.5	178.2	67.9	52
2	192	bmw	gas	std	two	sedan	rwd	front	101.2	176.8	64.8	54.
0	192	bmw	gas	std	four	sedan	rwd	front	101.2	176.8	64.8	54.
0	188	bmw	gas	std	two	sedan	rwd	front	101.2	176.8	64.8	54.
0	188	bmw	gas	std	four	sedan	rwd	front	101.2	176.8	64.8	54.
1		bmw	gas	std	four	sedan	rwd	front	103.5	189	66.9	55.
0		bmw	gas	std	four	sedan	rwd	front	103.5	189	66.9	55.
0		bmw	gas	std	two	sedan	rwd	front	103.5	193.8	67.9	53.
0		bmw	gas	std	four	sedan	rwd	front	110	197	70.9	56.
2	121	chevrolet	gas	std	two	hatchback	fwd	front	88.4	141.1	60.3	53.

view as

› Statistics

Visualizations

To create a graph, select a column in the table

Figure 3-2. *Infographic of data, types, and distributions*

Let's select a column to learn more about that feature. If you select the make column, information about that column is presented on the right side of the screen. You can see that there are 22 unique values for this feature, there are no missing values, and the feature is a string data type. Below this information is a histogram that displays the frequency of values of this feature; there are 32 instances of the value Toyota and only 9 instances of the value Dodge, as illustrated in Figure 3-3.

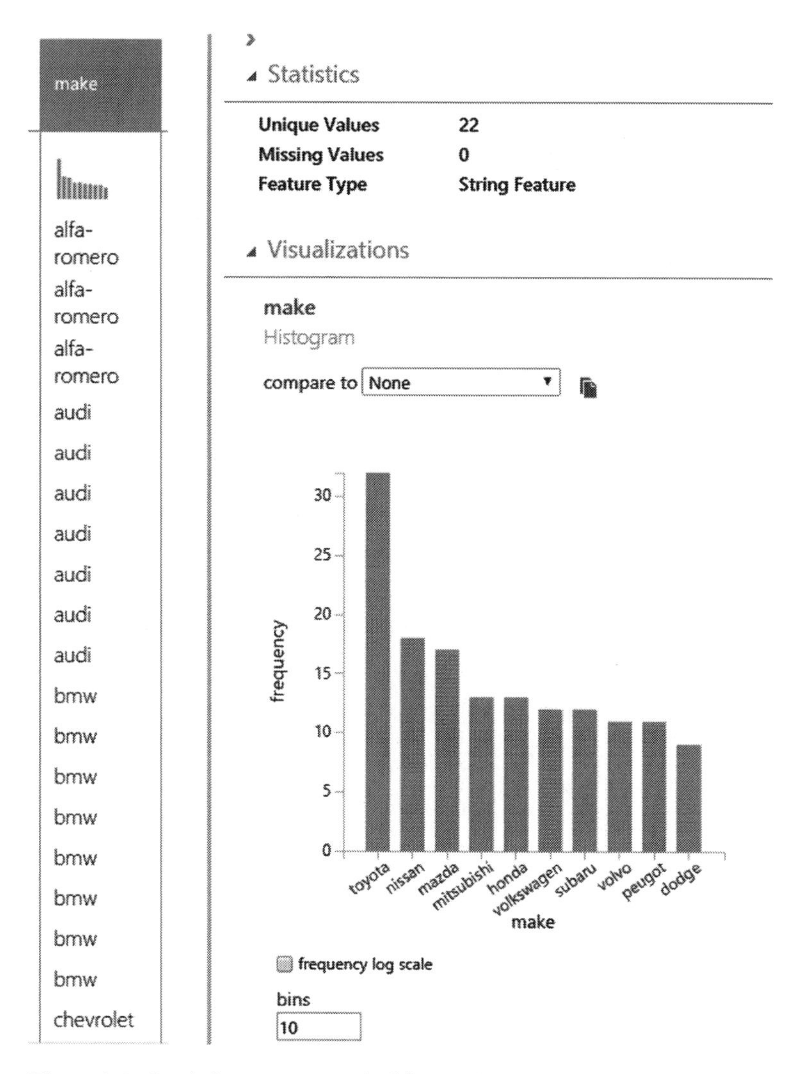

Figure 3-3. *Statistics on a categorical feature in Azure ML*

If you select a numeric feature in the dataset, such as horsepower, the information in Figure 3-4 is displayed.

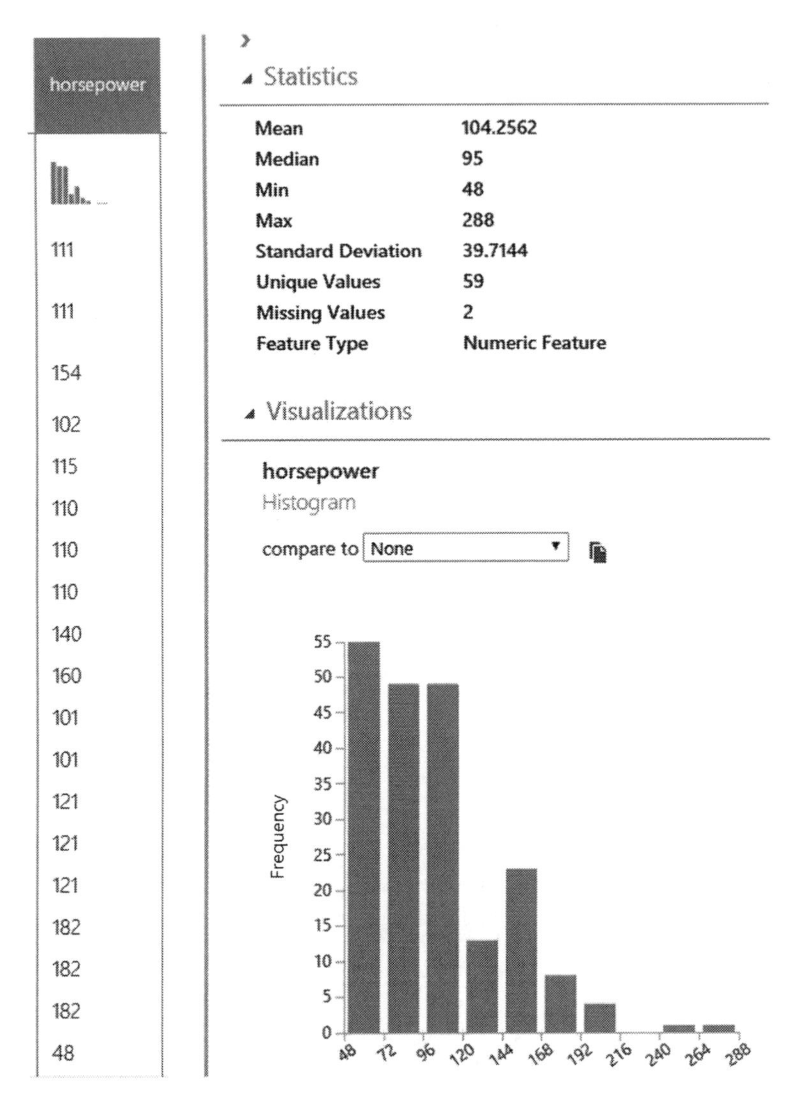

Figure 3-4. *Statistics on a numeric feature in Azure ML*

In the right hand side of Figure 3-4 are statistics on this numeric attribute, such as the mean, median, both min and max values, standard deviation, along with the number of unique values and number of missing values. And, as with a nominal attribute, you see the frequency of values for the attribute displayed; note that cars with 48 horsepower occur with the highest frequency–55 instances to be exact.

If you wish to learn even more information about your dataset, you can attach the Descriptive Statistics module, and once the experiment run is finished, you can visualize the results, as illustrated in Figure 3-5.

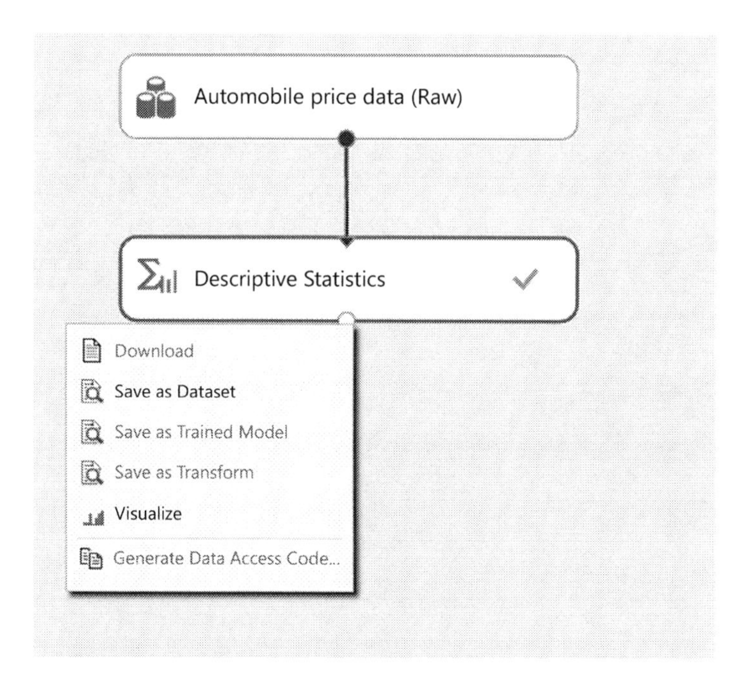

Figure 3-5. *Descriptive Statics module in Azure ML*

The Descriptive Statistics module creates standard statistical measures that describe each attribute in the dataset, illustrated in Figure 3-6. For each attribute, the module generates a row that begins with the column name and is followed by relevant statistics for the column based on its data type. You can use this report to quickly understand the features of the complete dataset, ranging from the number of rows in which values are missing, and the number of unique categorical values, to the mean and standard deviation of the column, and much more. But this visualization is merely a partial list of the statistics computed by the Descriptive Statistics module. To get the complete list, you merely need to save the dataset generated by this module and, instead of selecting the visualize option, convert to CSV using an Azure ML module. Then you can use Power BI or other reporting tool to fully explore the results.

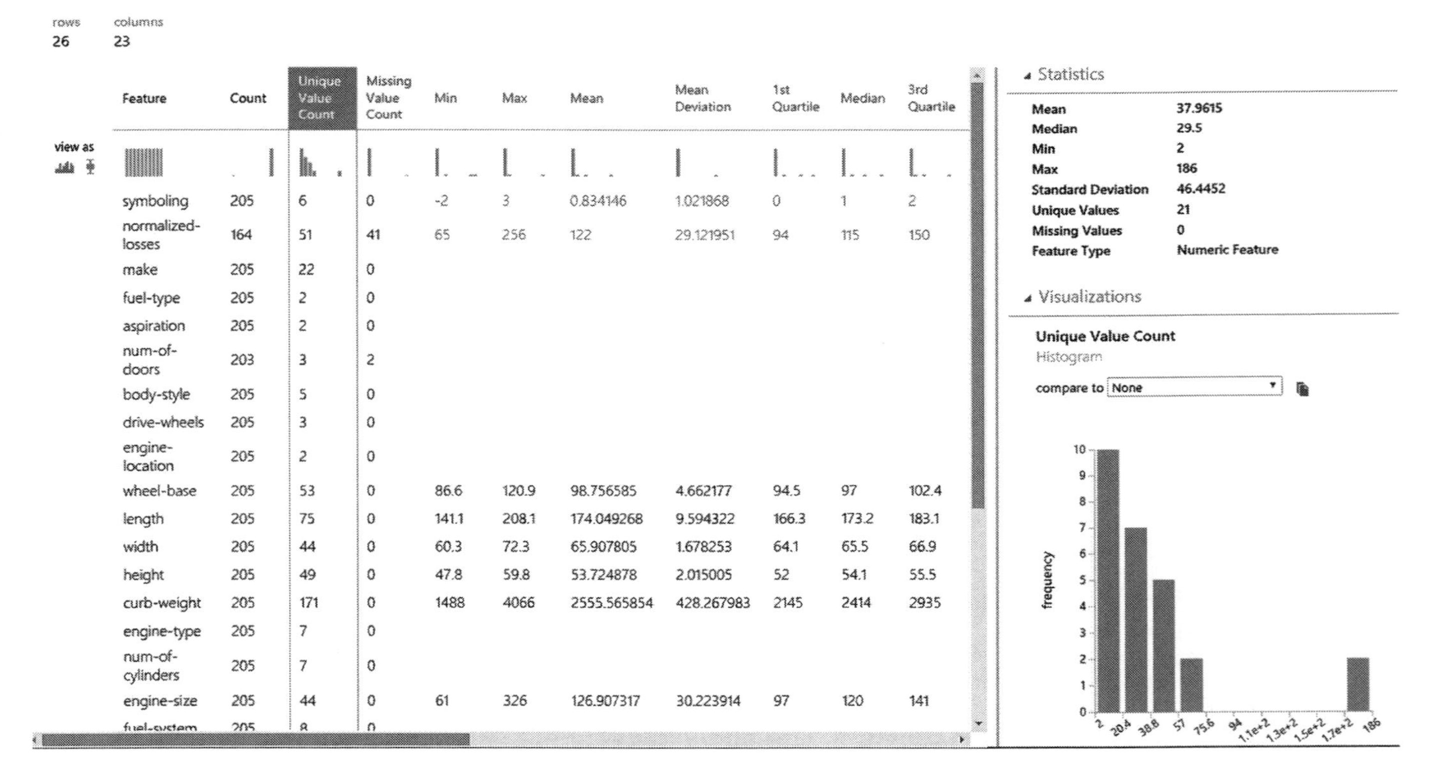

Figure 3-6. *Display of descriptive statics created by Azure ML*

When you find issues in your data, processing steps are necessary, such as cleaning missing values, removing duplicate rows, data normalization, discretization, dealing with mixed data types in a fields, and others. Data cleaning is a time-consuming task, and you will likely spend most of your time preparing your data, but it is absolutely necessary to build a high-quality model. The major data processing steps that we will cover in the remainder of this section are

- Handling missing and null values
- Removing duplicate records
- Identifying and removing outliers
- Feature normalization
- Dealing with class imbalance

Missing and Null Values

Many datasets encountered in practice contain missing values. Keep in mind that the data you are using to build a predictive model using Azure ML was almost certainly not created specifically for this purpose. When originally collected, many of the fields probably didn't matter and were left blank or unchecked. Provided it does not affect the original purpose of the data, there was no incentive to correct this situation; you, however, have a strong incentive to remove missing values from the dataset.

Missing values are frequently indicated by out-of-range entries, such as a negative number in a numeric attribute that is normally only positive, or a 0 in a numeric field that can never normally be 0. For nominal attributes, missing values may be simply indicated by blanks. You have to carefully consider the significance of missing values. They may occur for a number of reasons. Respondents in a survey may refuse to answer certain questions about their income, perhaps a device being monitored failed and it was impossible to obtain additional measurements, or the customer being tracked churned and terminated their subscription. What do these things mean about the example under consideration? Is the value for an attribute randomly missing or does the fact a value is missing carry information relevant to the task at hand, which we might wish to encode in the data?

Most machine learning schemes make the implicit assumption that there is no particular significance in the fact that a certain instance has an attribute value missing; the value is simply not known. However, there may be a good reason why the attribute's value is unknown—perhaps a decision was taken, on the evidence available, not to perform some particular test—and that might convey some information about the instance other than the fact that the value is simply missing. If this is the case, it is more appropriate to record it as "not tested" or encode a special value for this attribute or create another attribute in the dataset to record the value was missing. You need to dig deeper and understand how the data was generated to make an informed judgment about if a missing value has significance or whether it should simply be coded as an ordinary missing value.

Regardless of why a value is missing, this is a data quality issue that can cause problems downstream in your experiment and typically must be handled. You handle the missing or null value by either removing the offending row or by substituting in a value for the missing values. In either case, the Azure Machine Learning Clean Missing Values module is your tool of choice for cleaning missing values from your dataset. To configure the Clean Missing Values module, shown in Figure 3-7, you must first decide how to handle the missing values. This module supports both the removal and multiple replacement scenarios. The options are to

- Replace missing values with a placeholder value that you specify. This is useful when you wish to explicitly note missing values when they may contain useful information for modeling.
- Replace missing values with a calculated value, such as a mean, median, or mode of the column, or replace the missing value with an imputed value using multivariate imputation using chained equations (MICE) method or replace a missing value with Probabilistic PCA.
- Remove rows or columns that contain missing values, or that are completely empty.

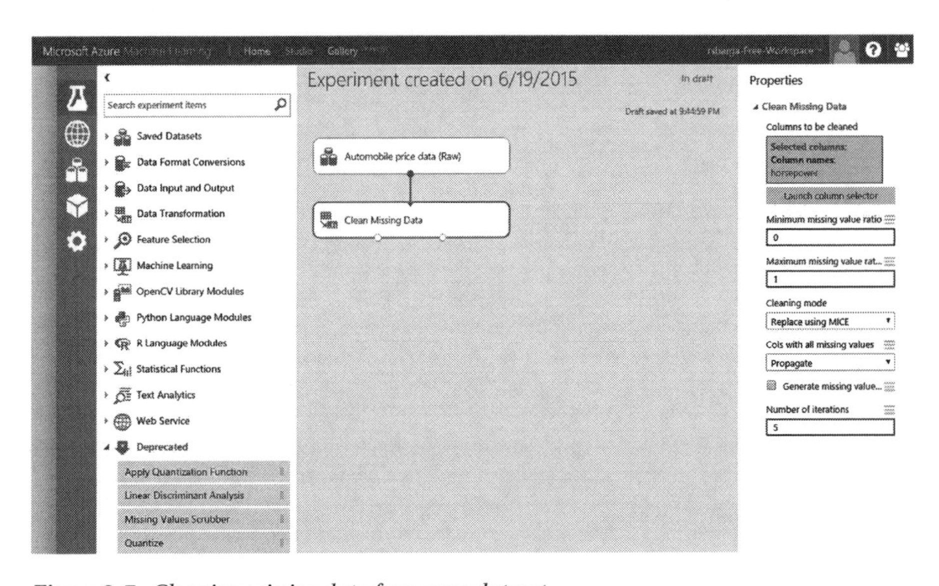

Figure 3-7. Cleaning missing data from your dataset

You will continue using the automobile price data, which has missing values in the horsepower attribute, and replace these missing values using the Clean Missing Data module set to replace missing values using the MICE method.

The Clean Missing Data module provides other options for replacing missing values, as illustrated in Figure 3-8, which are worth noting. In the first example, missing values are replaced with a custom substitution value, a fixed user-defined value, which is zero(0) in this example, which could be used to flag these values as missing. In the second example, if you assume the values are missing at random, you can replace them with median value for that column. In the third example, missing values are replaced using Probabilistic PCA, which computes replacement values based on other feature values within that same row and overall dataset statistics. You also generate a new feature (column) that indicates which columns were missing and filled in by the Clean Missing Data module.

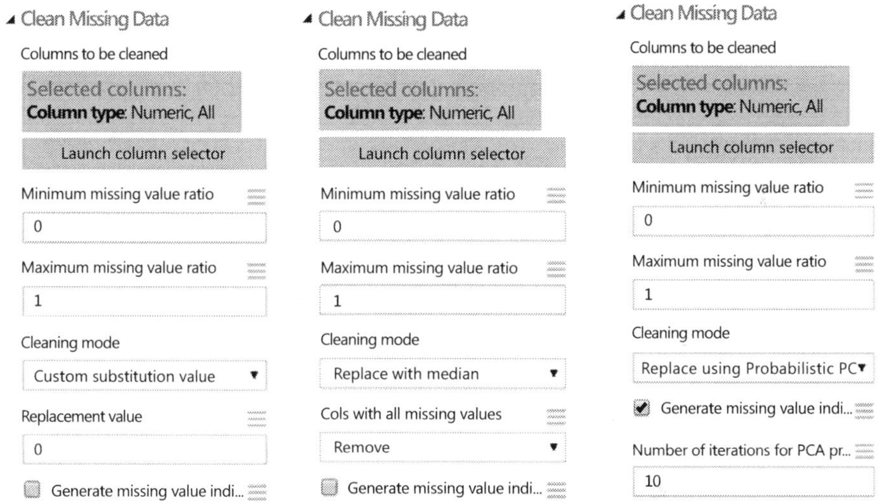

***Figure 3-8.** Options for replacing missing values in a dataset*

Handling Duplicate Records

Duplicate data presents another source of error that, if left unhandled, can impair the performance of the model you build. Most machine learning tools will produce different results if some of the instances in the data files are duplicated because repetition gives them more influence on the result. To identify and handle these records, you will identify and remove the duplicate rows using the Remove Duplicate Rows module in ML Studio. This module takes your dataset as input with uniqueness defined by either a single or combination of columns, which you identify using the Column Selector. The module will remove all duplicate rows from the input data, where two rows are considered duplicates if the values of all selected attributes are equal. It's noteworthy that NULLs are treated differently than empty strings, so you will need to take that into consideration if your dataset contains both. The output of the task is that your datasets have duplicate records removed.

For example, in the Movie Recommendation model that is provided in Azure ML, along with the IMDB Movie Titles dataset available in Azure ML, you need to remove all duplicates so that you have only one rating per each user per movie. The Remove Duplicate Rows module, illustrated in Figure 3-9, makes quick work of this and returns a dataset with only unique entries.

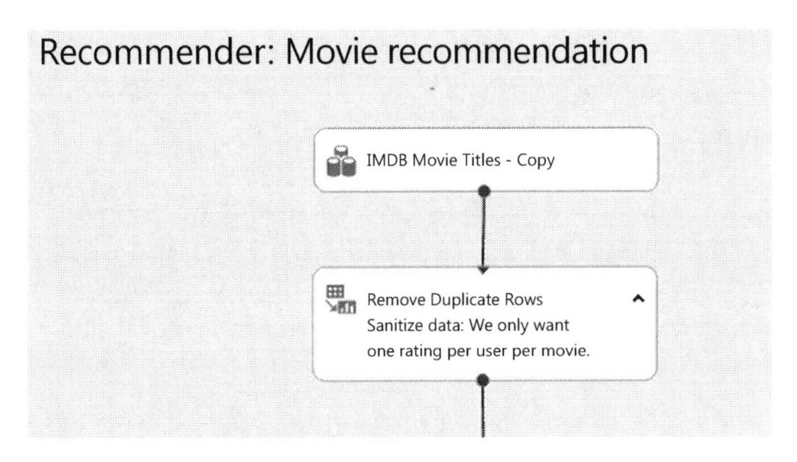

Figure 3-9. *Removing duplicate rows from the IMDB dataset*

Identifying and Removing Outliers

It is important to check your dataset for values that are far out of the normal range for the attribute. Outliers in a datasets are measurements that are significantly distant from other observations and can be a result of variability or even data entry error. Typographical or measurement errors in numeric values generally create outliers that can be detected by graphing one variable at a time using the visualize option for the dataset and selecting the box plot option for each column, as illustrated in Figure 3-10. Outliers deviate significantly from the pattern that is apparent in the remaining values.

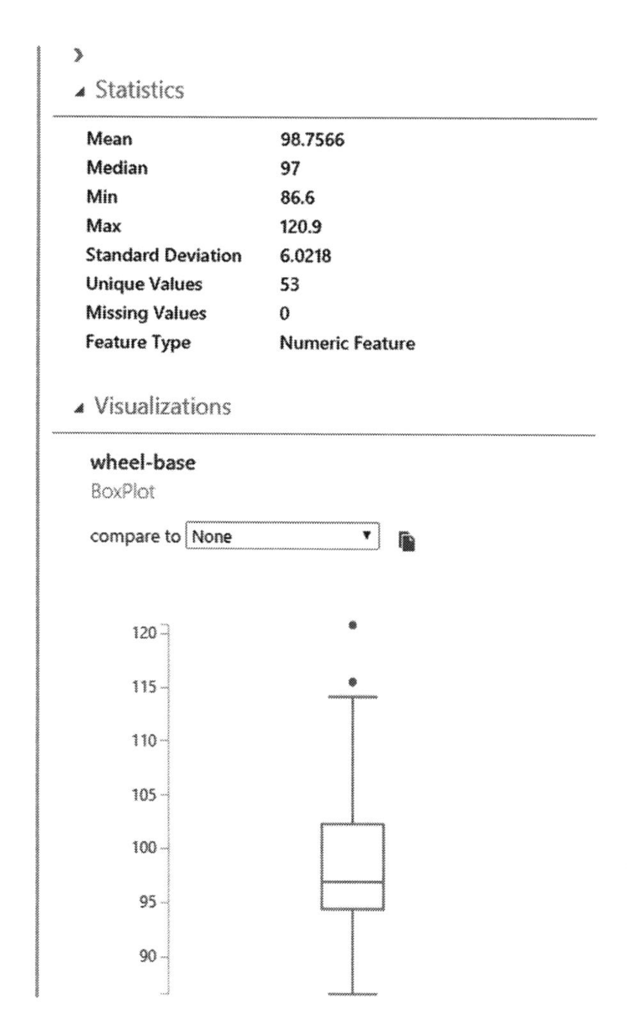

Figure 3-10. *Visualizing data distributions using box plots*

Often, outliers are hard to find through manual exploration, particularly without specialist domain knowledge. When not handled, outliers can skew the results of your experiments, leading to suboptimal results. To remove outliers or clip them, you will use the Clip Values module in ML Studio, illustrated in Figure 3-11.

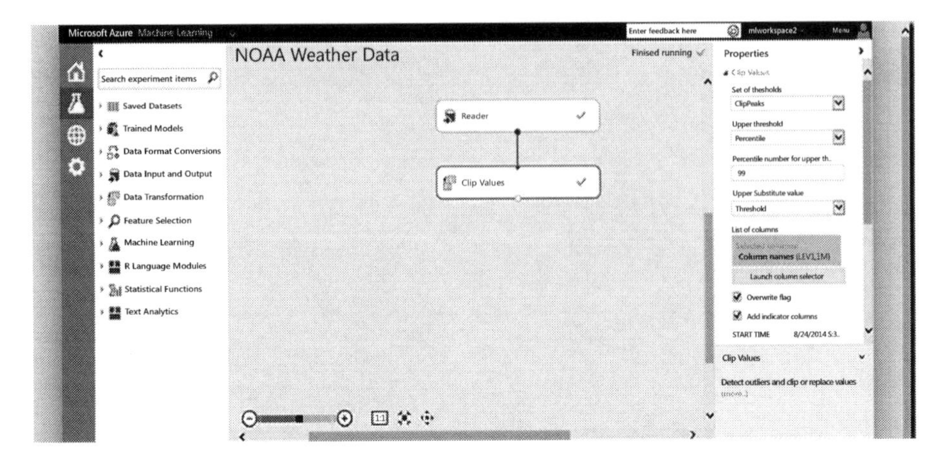

Figure 3-11. *Clipping values using the Clip Values module*

The Clip Values module accepts a dataset as input and is capable of clipping data point values that exceed a specified threshold using either a specified constant or percentile for the selected columns. The outliers, both peak and sub-peak, can be replaced with the threshold, mean, median, or a missing value. Optionally, a column can be added to indicate whether a value was clipped or not. The Clip Values module expects columns containing numeric data. This module constructs a new column with the peak values clipped to the desired threshold and a column that indicates the values that were clipped. Optionally, clipped values can be written to the original column in place. There are multiple ways to identify what constitutes an outlier and you must specify which method you wish to use:

- ClipPeak looks for and then clips or replaces values that are greater than a specified upper boundary value.
- ClipSubpeaks looks for and then clips or replaces values less than a specified lower boundary value.
- ClipPeaksAndSubpeaks applies both a lower and upper boundary on values, and clips or replaces all values outside that range.

You set the values to use as the upper or lower boundary, or you can specify a percentile range. All values outside the specified percentiles are replaced or clipped using the method you specify. Missing values are not clipped when they are propagated to the output dataset, and missing values are ignored when mean or median is computed for a column. The column indicating clipped values contains False for them. The Clip Values module allows you to target specific columns by selecting which columns the Clip Values modules will work with. This allows you to use different strategies for identifying outliers and different replacement strategies on a per column basis.

Feature Normalization

Once you have replaced missing values in the dataset, removed outliers, and pruned redundant rows, it is often useful to normalize the data so that it is consistent in some way. If your dataset contains attributes that are on very different scales, such as a mix of measurements in millimeters, centimeters, and meters, this can potentially add error or distort your experiment when you combine the values as features during modeling. By transforming the values so that they are on a common scale, yet maintain their general distribution and ratios, you can generally get better results when modeling. To perform feature normalization, you need to go through each feature in your dataset and adjust it the same way across all examples so that the columns in the dataset are on a common scale. To do this, you will use the Normalize Data module in Azure ML Studio.

The Normalize Data module offers five different mathematical techniques to normalize attributes and it can operate on one or more selected columns. To transform a numeric column using a normalization function, select the columns using Column Selector, then select from the following functions:

- **Zscore:** This option converts all values to a z-score.
- **MinMax:** This option linearly rescales every feature to the [0,1] interval by shifting the values of each feature so that the minimal value is 0, and then dividing by the new maximal value (the difference between the original maximal and minimal values).
- **Logistic:** The values in the column are transformed using a log transform.
- **LogNormal:** This option converts all values to a lognormal scale.
- **Tanh:** All values are converted to a hyperbolic tangent.

To illustrate, you will use the German Credit Card dataset from the UC Irvine repository, available in ML Studio. This dataset contains 1,000 samples with 20 features and 1 label. Each sample represents a person. The 20 features include both numerical and categorical features. The last column is the label, which denotes credit risk and has only two possible values: high credit risk = 2, and low credit risk = 1. The machine learning algorithm you will use requires that data be normalized. Therefore, you use the Normalize Data module to normalize the ranges of all numeric features, using a tanh transformation, illustrated in Figure 3-12. A tanh transformation converts all numeric features to values within a range of 0-1, while preserving the overall distribution of values.

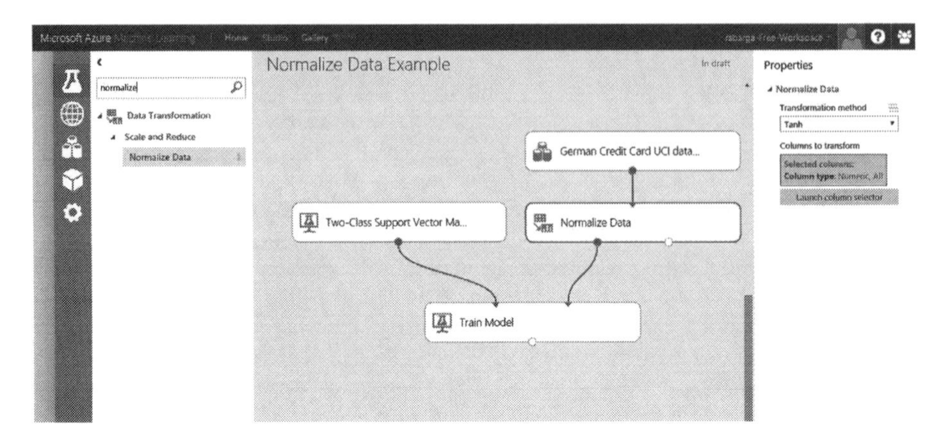

Figure 3-12. *Normalizing dataset with the Tanh transformation*

Dealing with Class Imbalance

Class imbalance in machine learning is where the total number of a class of data (class 1) is far less than the total number of another class of data (class 2). This problem is extremely common in practice and can be observed in various disciplines including fraud detection, anomaly detection, medical diagnosis, oil spillage detection, and facial recognition. Most machine learning algorithms work best when the number of instances of each classes are roughly equal. When the number of instances of one class far exceeds the other, it usually produces a biased classifier that has a higher predictive accuracy over the majority class(es), but poorer predictive accuracy over the minority class. SMOTE (Synthetic Minority Over-sampling Technique) is specifically designed for learning from imbalanced datasets.

SMOTE is one of the most adopted approaches to deal with class imbalance due to its simplicity and effectiveness. It is a combination of oversampling and undersampling. The majority class is first under-sampled by randomly removing samples from the majority class population until the minority class becomes some specified percentage of the majority class. Oversampling is then used to build up the minority class, not by simply replicating data instances of the minority class but, instead, it constructs new minority class data instances via an algorithm. SMOTE creates synthetic instances of the minority class. By synthetically generating more instances of the minority class, inductive learners are able to broaden their decision regions for the minority class.

Azure ML includes a SMOTE module that you can apply to an input dataset prior to machine learning. The module returns a dataset that contains the original samples, plus an additional number of synthetic minority samples, depending on a percentage that you specify. As previously described, these new instances are created by taking samples of the feature space for each target class and its nearest neighbors, and generating new examples that combine features of the target case with features of its neighbors. This approach increases the features available to each class and makes the samples more general. To configure the SMOTE module, you simply specify the number of nearest neighbors to use in building feature space for new cases and the percentage of the minority class to increase (100 increases the minority cases by 100%, 200 increases the minority cases by 200%, etc.). You use the Metadata Editor module to select the column that contains the class labels, using select Label from the Fields drop-down list. Based on this metadata, SMOTE identifies the minority class in the label column to get all examples for the minority class. SMOTE never changes the number of majority cases.

For illustrative purposes, you will use the blood donation dataset available in Azure ML Studio. While this dataset is not very imbalanced, it serves the purpose for this example. If you visualize the dataset, you can see that of the 748 examples in the dataset, there are 570 cases (76%) of Class 0 and only 178 cases (24%) of class 1. Class 1 represents the people who donated blood, and thus these rows contain the features you want to model. To increase the number of cases, you will use the SMOTE module, as illustrated in Figure 3-13.

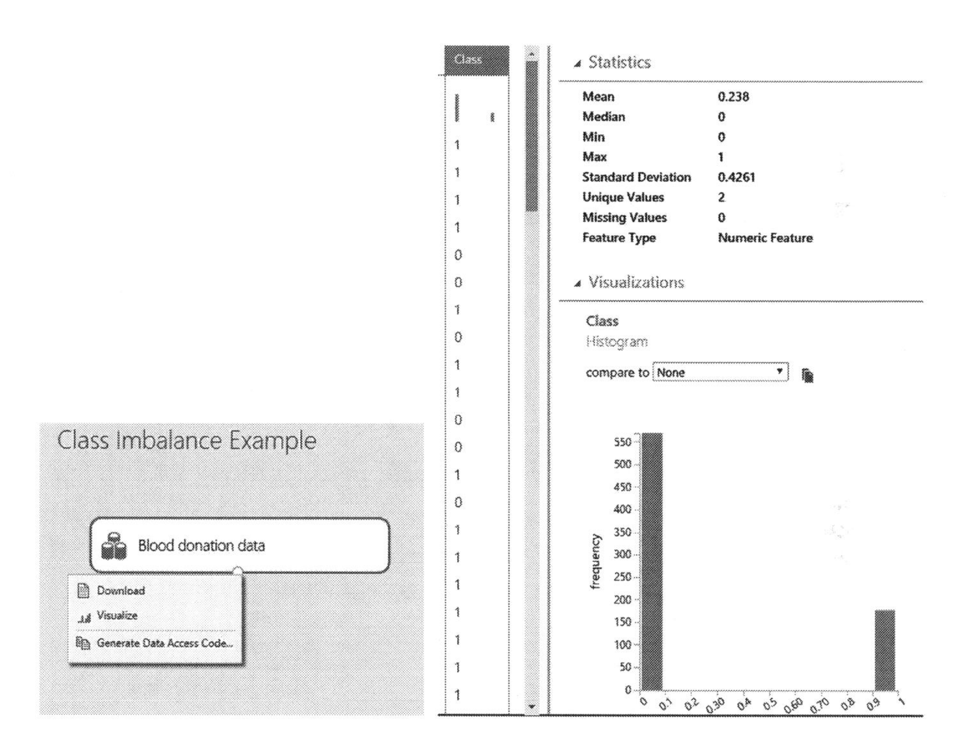

Figure 3-13. *Visualizing the class imbalance in the blood donation dataset, which is available in Azure ML*

You use the Metadata Editor module to identify the class label and then send the dataset to the SMOTE module and set the percentage increase to 100 and nearest neighbors to 2. Smote will select a data instance from the minority class, find the two nearest neighbors to that instance, and use a blend of values from the instance and its two nearest neighbors to create a new synthetic data instance. Once the experiment is complete, you visualize the dataset, illustrated in Figure 3-14; note that the minority class has increased while the majority class remains unchanged.

Figure 3-14. Visualizing the class distribution in the blood donation dataset after running the SMOTE module

Feature Selection

In practice, adding irrelevant or distracting attributes to a dataset often confuses machine learning systems. Feature selection is the process of determining the features with the highest information value to the model. The two main approaches are the filtering and wrapper methods. Filtering methods analyze features using a test statistic and eliminate redundant or non-informative features. As an example, a filtering method could eliminate features that have little correlation to the class labels. Wrapper methods utilize a classification model as part of feature selection. One example of the wrapper method is the decision tree, which selects the most promising attribute to split on at each point and should, in theory, never select irrelevant or unhelpful attributes.

There are tradeoffs between these techniques. Filtering methods are faster to compute since each feature only needs to be compared against its class label. Wrapper methods, on the other hand, evaluate feature sets by constructing models and measuring performance. This requires a large number of models to be trained and evaluated (a quantity that grows exponentially in the number of features). With filter methods, a feature with weak correlation to its class labels is eliminated. Some of these eliminated features, however, may have performed well when combined with other features.

You should find an approach that works best for you, incorporate it into your data science workflow, and run experiments to determine its effectiveness. You can worry about optimization and tuning your approaches later during incremental improvement.

In this section, we will describe how to perform feature selection in Azure ML Studio. Feature selection can increase model accuracy by eliminating irrelevant and highly correlated features, which in turn will make model training more efficient. While feature selection seeks to reduce the number of features in a dataset, it is not usually referred to by the term "dimensionality reduction". Feature selection methods extract a subset of original features in the data without changing them. Dimensionality reduction methods create engineered features by transforming the original features. We will examine dimensionality reduction in the following section of this chapter.

In Azure Machine Learning Studio, you will use the Filter-Based Feature Selection module to identify the subset of input columns that have the greatest predictive power. The module, illustrated in Figure 3-15, outputs a dataset containing the specified number of best feature columns, as ranked by predictive power, and also outputs the names of the features and their scores from the selected metric.

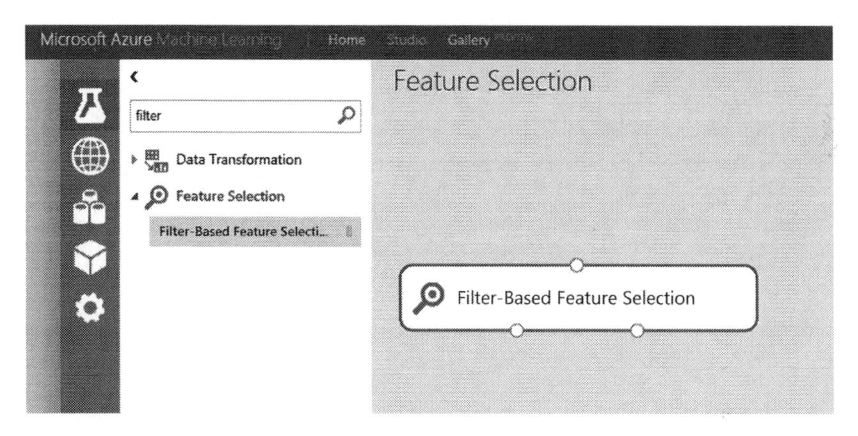

Figure 3-15. *The Filter-Based Feature Selection module*

The Filter-Based Feature Selection module provides multiple feature selection algorithms, which you apply based on the type of predictive task and data types. Evaluating the correlation between each feature and the target attribute, these methods apply a statistical measure to assign a score to each feature. The features are then ranked by the score, which may be used to help set the threshold for keeping or eliminating a specific feature. The following is the list of widely used statistical tests in the Filter-Based Feature Selection module for determining the subset of input columns that have the greatest predictive power:

- **Pearson Correlation:** Pearson's correlation statistics, or Pearson's correlation coefficient, is also known in statistical models as the r value. For any two variables, it returns a value that indicates the strength of the correlation. Pearson's correlation coefficient is computed by taking the covariance of two variables and dividing by the product of their standard deviations. The coefficient is not affected by changes of scale in the two variables.

- **Mutual Information:** The Mutual Information Score method measures the contribution of a variable towards reducing uncertainty about the value of another variable —in this case, the label. Many variations of the mutual information score have been devised to suit different distributions. The mutual information score is particularly useful in feature selection because it maximizes the mutual information between the joint distribution and target variables in datasets with many dimensions.
- **Kendall Correlation:** Kendall's rank correlation is one of several statistics that measure the relationship between rankings of different ordinal variables or different rankings of the same variable. In other words, it measures the similarity of orderings when ranked by the quantities. Both this coefficient and Spearman's correlation coefficient are designed for use with non-parametric and non-normally distributed data.
- **Spearman Correlation:** Spearman's coefficient is a nonparametric measure of statistical dependence between two variables, and is sometimes denoted by the Greek letter rho. The Spearman's coefficient expresses the degree to which two variables are monotonically related. It is also called Spearman rank correlation because it can be used with ordinal variables.
- **Chi-Squared:** The two-way chi-squared test is a statistical method that measures how close expected values are to actual results. The method assumes that variables are random and drawn from an adequate sample of independent variables. The resulting chi-squared statistic indicates how far results are from the expected (random) result.
- **Fisher Score:** The Fisher score (also called the Fisher method, or Fisher combined probability score) is sometimes termed the information score because it represents the amount of information that one variable provides about some unknown parameter on which it depends. The score is computed by measuring the variance between the expected value of the information and the observed value. When variance is minimized, information is maximized. Since the expectation of the score is zero, the Fisher information is also the variance of the score.
- **Count-Based:** Count-based feature selection is a simple yet relatively powerful way of finding information about predictors. It is a non-supervised method of feature selection, meaning you don't need a label column. This method counts the frequencies of all values and then assigns a score to the column based on frequency count. It can be used to find the weight of information in a particular feature and reduce the dimensionality of the data without losing information.

Consider, for example, the use of the Filter-Based Feature Selection module. For illustration, you will use the Pima Indians Diabetes Binary Classification dataset available in ML Studio (shown in Figure 3-16). There are eight numeric features in this dataset, along with the class label, and you will reduce this to the top three features.

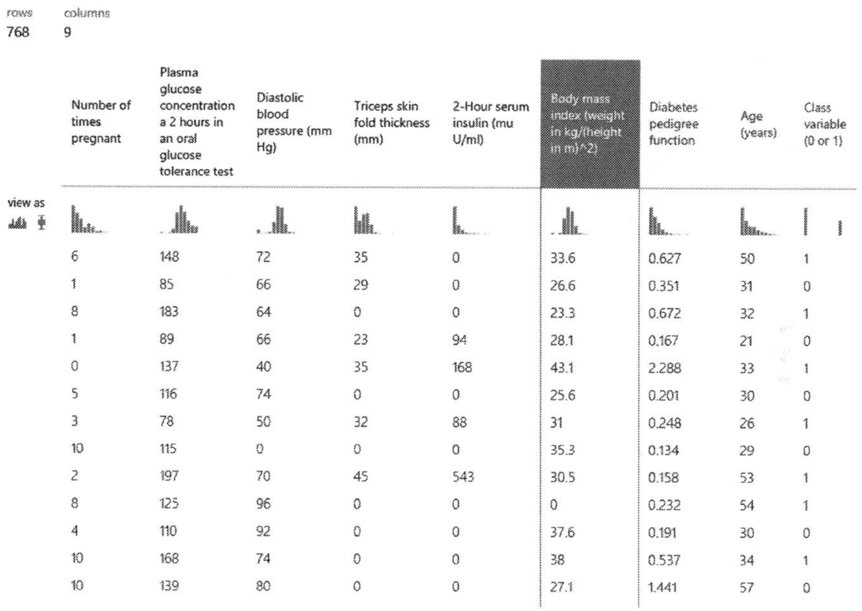

rows 768
columns 9

view as

Number of times pregnant	Plasma glucose concentration a 2 hours in an oral glucose tolerance test	Diastolic blood pressure (mm Hg)	Triceps skin fold thickness (mm)	2-Hour serum insulin (mu U/ml)	Body mass index (weight in kg/(height in m)^2)	Diabetes pedigree function	Age (years)	Class variable (0 or 1)
6	148	72	35	0	33.6	0.627	50	1
1	85	66	29	0	26.6	0.351	31	0
8	183	64	0	0	23.3	0.672	32	1
1	89	66	23	94	28.1	0.167	21	0
0	137	40	35	168	43.1	2.288	33	1
5	116	74	0	0	25.6	0.201	30	0
3	78	50	32	88	31	0.248	26	1
10	115	0	0	0	35.3	0.134	29	0
2	197	70	45	543	30.5	0.158	53	1
8	125	96	0	0	0	0.232	54	1
4	110	92	0	0	37.6	0.191	30	0
10	168	74	0	0	38	0.537	34	1
10	139	80	0	0	27.1	1.441	57	0

Figure 3-16. *Pima Indians Diabetes Binary Classification data*

Set **Feature scoring method** to be **Pearson Correlation**, the **Target column** to be **Col9**, and the **Number of desired features** to **3**. Then the module Filter-Based Feature Selection will produce a dataset containing three features together with the target attribute called Class Variable. Figure 3-17 shows the flow of this experiment and the input parameters just described.

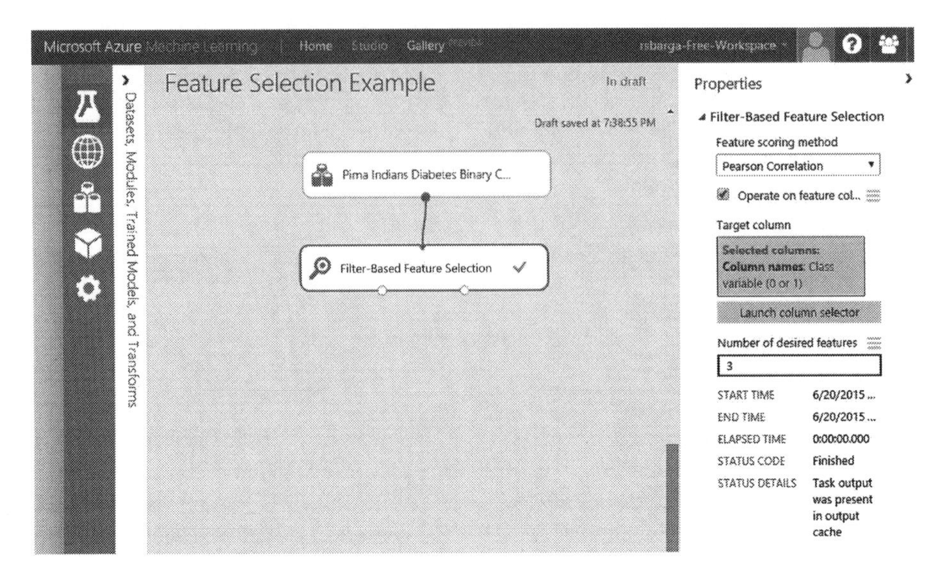

***Figure 3-17.** Pearson Correlation's top three features*

Figure 3-18 shows the resulting dataset. Each feature is scored based on the Pearson Correlation with the target attribute. Features with top scores are kept, namely 1) plasma glucose, 2) body mass index, and 3) age.

Feature Selection Example › Filter-Based Feature Selection › Filtered dataset

rows 768, columns 4

Class variable (0 or 1)	Plasma glucose concentration a 2 hours in an oral glucose tolerance test	Body mass index (weight in kg/(height in m)^2)	Age (years)
1	148	33.6	50
0	85	26.6	31
1	183	23.3	32
0	89	28.1	21
1	137	43.1	33
0	116	25.6	30
1	78	31	26
0	115	35.3	29
1	197	30.5	53
1	125	0	54

Filter-Based Feature Selection

Right click the first port

***Figure 3-18.** Top three features by Pearson Correlation*

The corresponding scores of the selected features are shown in Figure 3-19. Plasma glucose has a Pearson correlation of 0.66581, body mass a Pearson correlation of 0.292695, and age is 0.238356.

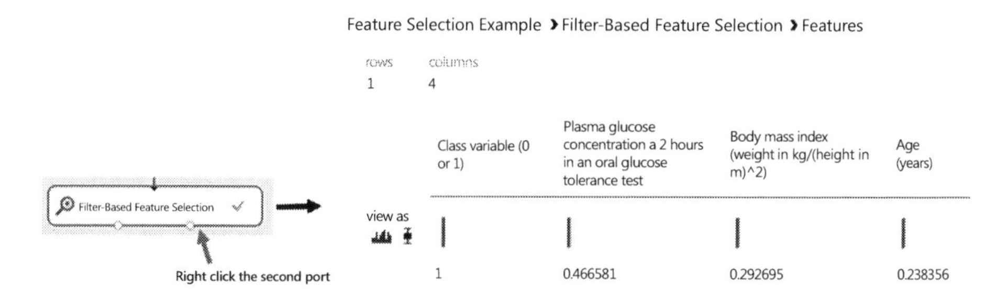

Figure 3-19. *Pearson Correlation feature correlation measures*

By applying this Filter Based Feature Selection module, three out of eight features are selected because they have the most correlated features with the target variable Class Variable, based on the scoring method Pearson Correlation.

Feature Engineering

Feature engineering is the process where you can establish the representation of data in the context of the modeling approach you have selected and imbue your domain expertise and understanding of the problem. It is the process of transforming raw data into features that better represent the underlying problem to the machine learning algorithm, resulting in improved model accuracy on unseen data. It is the foundational skill in data science. And, where running effective experiments is one aspect of the science in data science, then feature engineering is the art that draws upon your creativity and experience. If you are seeking inspiration, review the forum on Kaggle.com in which data scientists describe how they won Kaggle data science contests; it's a valuable source of information.

Feature engineering can be a complex task that may involve running experiments to try different approaches. Features may be simple, such as "bag of words," a popular technique in text processing, or may be based on your understanding of the domain. Knowledge of the domain in which a problem lies is immensely valuable and irreplaceable. It provides an in-depth understanding of your data and the factors influencing your analytic goal. Many times domain knowledge is a key differentiator to your success. Let's look at an illustrative example.

Assume you are attempting to build a model that will inform you whether a package should be sent via ground transportation or sent via air travel. Your training set consists of the latitude and longitude for the source and destination, illustrated in Figure 3-20, along with the classification for whether the package should be sent via ground transportation (drivable) or air travel (not drivable).

CITY 1 LAT.	CITY 1 LNG.	CITY 2 LAT.	CITY 2 LNG.	Driveable?
123.24	46.71	121.33	47.34	Yes
123.24	56.91	121.33	55.23	Yes
123.24	46.71	121.33	55.34	No
123.24	46.71	130.99	47.34	No

Figure 3-20. *Sample dataset to drive or fly a package*

It would be exceptionally difficult for a machine learning algorithm to learn the relationships between these four attributes and the class label. But even if the machine learning algorithm doesn't have knowledge of latitude and longitude, you do, so why not use this to engineer a feature that carries information the machine learning algorithm can use? If you simply compute the distance between the source and destination, as illustrated in Figure 3-21, and use this as the feature, any linear machine learning algorithm could learn this classification in short order.

DISTANCE (MI.)	Driveable?
14	Yes
28	Yes
705	No
2432	No

Figure 3-21. *A new feature, distance between destination*

Feature engineering is when you use your knowledge about the data to create fields that make a machine learning algorithm work better. Engineered features that enhance the training provide information that better differentiates the patterns in the data. You should strive to add a feature that provides additional information that is not clearly captured or easily apparent in the original or existing feature set. The modules in the Manipulation drawer of Azure ML, under the Data Transformation category, shown in Figure 3-22, will enable you to create these new features.

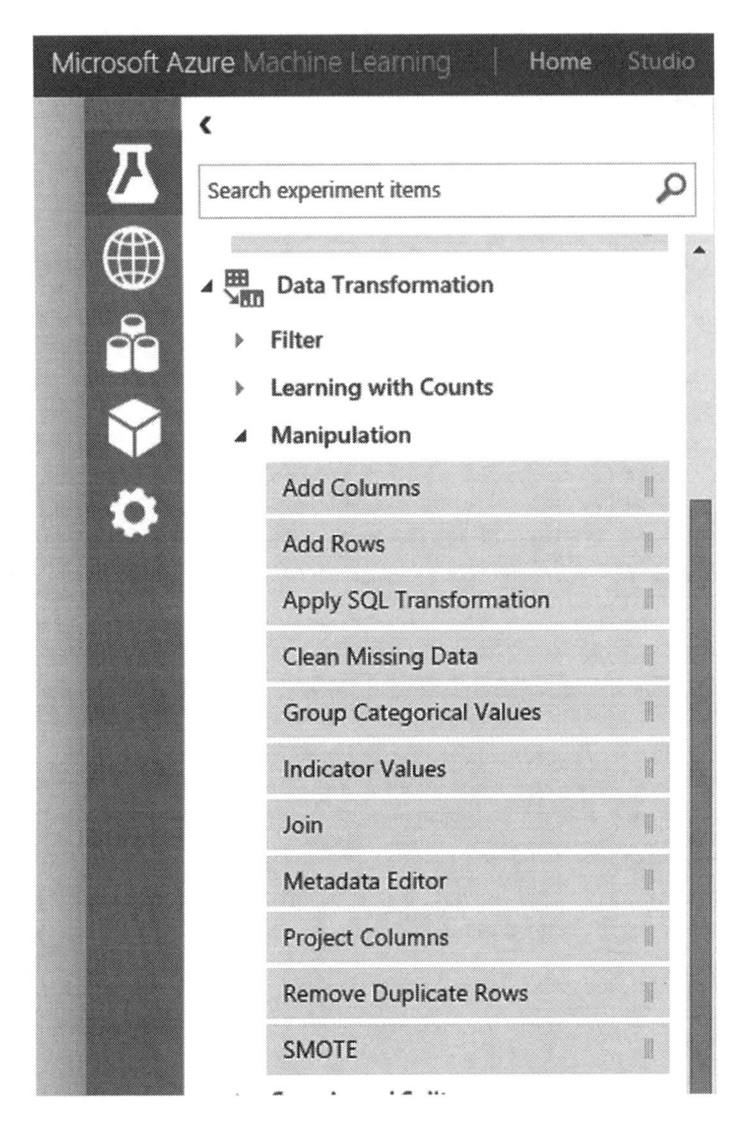

Figure 3-22. *Data manipulation modules in Azure ML*

Here are some considerations in feature engineering to help you get started:

- **Decompose Categorical Attributes:** Suppose you have a categorical attribute, like Item_Color, that can be Red, Blue, or Unknown. Unknown may be special, but to a model, it looks like just another color choice. It might be beneficial to better expose this information. You could create a new binary feature called Has_Color and assign it a value of 1 when an item has a color and 0 when the color is unknown. Going a step further, you could create a binary feature for each value that Item_Color has. This would be three binary attributes: Is_Red, Is_Blue, and Is_Unknown. These additional features could be used instead of the Item_Color feature (if you wanted to try a simpler linear model) or in addition to it (if you wanted to get more out of something like a decision tree).

- **Decompose Date-Time:** A date-time contains a lot of information that can be difficult for a model to take advantage of in its native form, such as ISO 8601 (i.e. 2014-09-20T20:45:40Z). If you suspect that there are relationships between times and other attributes, you can decompose a date-time into constituent parts that may allow models to discover and exploit these relationships. For example, you may suspect that there is a relationship between the time of day and other attributes. You could create a new numerical feature called Hour_of_Day for the hour that might help a regression model. You could create a new ordinal feature called Part_Of_Day with four values: 1) morning, 2) midday, 3) afternoon, and 4) night, with hour boundaries you think are relevant. You can use similar approaches to pick out time-of-week relationships, time-of-month relationships, and various structures of seasonality across a year.

- **Reframe Numerical Quantities:** Your data is very likely to contain quantities that can be reframed to better expose relevant structures. This may be a transform into a new unit or the decomposition of a rate into time and amount components. You may have a quantity like a weight, distance, or timing. A linear transform may be useful to regression and other scale dependent methods.

Great things happen in machine learning when human and machine work together, combining your knowledge of how to create relevant feature from the data with the machine learning algorithm talent for optimizing.

Binning Data

You may encounter a numeric feature in your dataset with a range of continuous numbers, with too many values to model; in visualizing the dataset, the first indication of this is the feature has a large number of unique values. In many cases, the relationship between such a numeric feature and the class label is not linear (the feature value does not increase or decrease monotonically with the class label). In such cases, you should consider binning the continuous values into groups using a technique known as quantization where each bin represents a different range of the numeric feature. Each categorical feature (bin) can then be modeled as having its own linear relationship with the class label. For example, say you know that the continuous numeric feature called mobile minutes used is not linearly correlated with the likelihood for a customer to churn. You can bin the numeric feature mobile minutes used into a categorical feature that might be able to capture the relationship with the class label (churn) more accurately. The optimum number of bins is dependent on characteristics of the variable and its relationship to the target, and this is best determined through experimentation.

The Quantize Data module in Azure ML Studio provides this capability out-of-the-box; it supports a user-defined number of buckets and provides multiple quantile normalization functions, which we describe below. In addition, the module can optionally append the bin assignment to the dataset, replacing the dataset value or returning a dataset that contains only the resulting bins.

During binning (or quantization), each value in a column is mapped to a bin by comparing its value against the values of bin edges. For example, if the value is 1.5 and the bin edges are 1, 2, and 3, the element would be mapped to bin number 2. Value 0.5 would be mapped to bin number 1 (the underflow bin), and value 3.5 would be mapped to bin number 4 (the overflow bin).

Azure Machine Learning provides several different algorithms for determining the bin edges and assigning numbers to bins:

- **Entropy MDL:** The entropy model for binning calculates an information score that indicates how useful a particular bin range might be in the predictive model.
- **Quantiles:** The quantile method assigns values to bins based on percentile ranks.
- **Equal Width:** With this option, you specify the total number of bins. The values from the data column will be placed in the bins such that each bin has the same interval between start and end values. As a result, some bins might have more values if data is clumped around a certain point.
- **Custom Edges:** You can specify the values that begin each bin. The edge value is always the lower boundary of the bin.
- **Equal Width with Custom Start and Stop:** This method is like the Equal Width option, but you can specify both lower and upper bin boundaries.

Because there are so many options, each customizable, you need to experiment with the different methods and values to identity the optimal binning strategy for your data.

For this example, you will use the public dataset Forest Fires Data, provided by the UCI Machine Learning repository and available in Azure ML Studio. The dataset is intended to be used to predict the burn area of forest fires in the northeast region of Portugal by using meteorological and other data. One feature in this dataset is DC, a drought code index, which ranges from 7.9 to 860.6 and there are a large number of unique values. You will bin the DC values into five bins.

Pull the dataset into your experiment; pull the Quantize Data module onto the pallete; connect the dataset to the module; click the Quantize Data module; select the quantiles binning mode and specify five bins, identify the DC column as the target for binning; and select the option to add this new binned feature to the dataset, as illustrated in Figure 3-23.

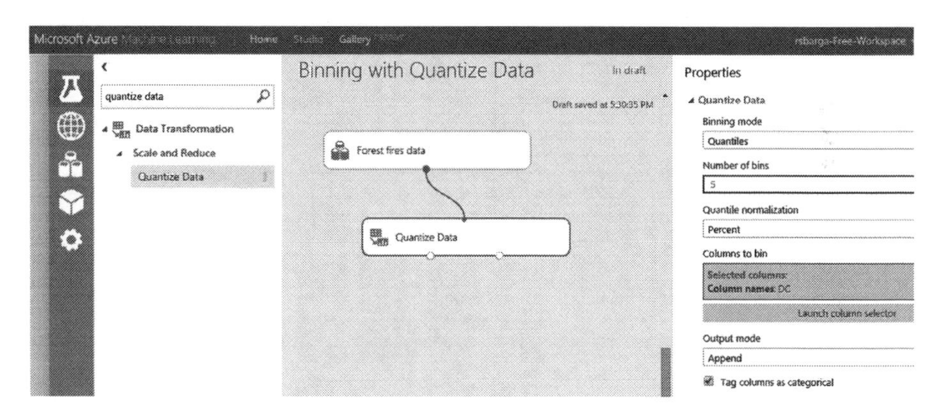

***Figure 3-23.** Quantize Data module in Azure ML*

Once complete, you can visualize the resulting dataset, illustrated in Figure 3-24, and scroll to the newly created column `DC_quantized` and select that column, at which point you can see that you have created a new categorical variable with five unique values corresponding to the bins.

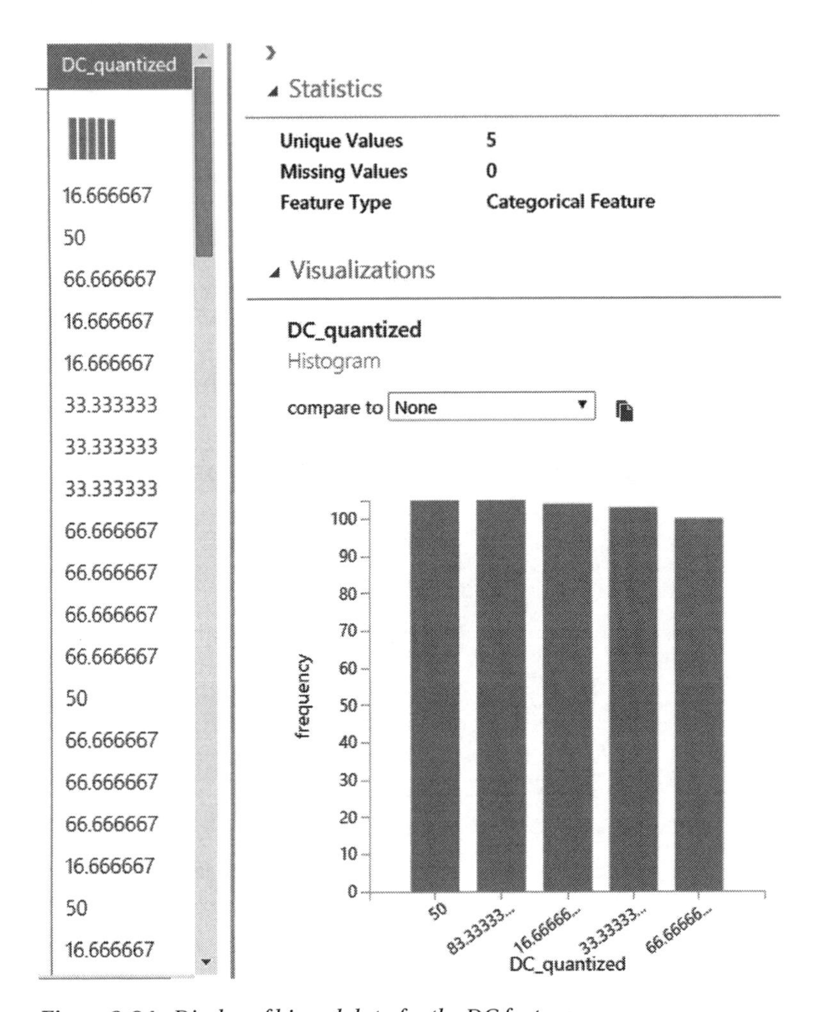

Figure 3-24. *Display of binned data for the DC feature*

You can apply a different quantization rule to different columns by chaining together multiple instances of the Quantize Data module, and in each instance, selecting a subset of columns to quantize.

If the column to bin (quantize) is sparse, then the bin index offset (quantile offset) is used when the resulting column is populated. The offset is chosen so that sparse 0 always goes to bin with index 0 (quantile with value 0). As a result, sparse zeros are propagated from the input to output column. All NaNs and missing values are propagated from the input to output column. The only exception is the case when the module returns quantile indexes, in which case all NaNs are promoted to missing values.

The Curse of Dimensionality

The "curse of dimensionality" is one of the more important results in machine learning. Much has been written on this phenomenon, but it can take years of practice to appreciate its true implications. Classification methods are subject to the implications of the curse of dimensionality. The basic intuition is that as the number of data dimensions increases, it becomes more difficult to create a model that generalizes and applies well to data not in the training set. This difficulty is often hard to overcome in real-world settings. As a result, you must work to minimize the number of dimensions. This requires a combination of clever feature engineering and use of dimensionality reduction techniques. There is no magical potion to cure the curse of dimensionality, but there is Principal Components Analysis (PCA).

In many real applications, you will be confronted with various types of high-dimensional data. The goal of dimensionality reduction is to convert this data into a lower dimensional representation, such that uninformative variance in the data is discarded and the lower dimensional data is more amenable to processing by machine learning algorithms. PCA is a statistical procedure that uses a mathematical transformation to convert your training dataset into a set of values of linearly uncorrelated variables called principal components. The number of principal components is less than or equal to the number of original attributes in your training set. The transformation is defined in such a way that the first principal component has the largest possible variance and accounts for as much of the variability in the data as possible, and each succeeding component in turn has the highest variance possible under the constraint that it is orthogonal to (uncorrelated with) the preceding components. The principal components are orthogonal because they are the eigenvectors of the covariance matrix, which is symmetric. PCA is sensitive to the relative scaling of the original variables, so your dataset must be normalized prior to PCA for good results. You can view PCA as a data-mining technique. The high-dimensional data is replaced by its projection onto the most important axes. These axes are the ones corresponding to the largest eigenvalues. Thus, the original data is approximated by data that has many fewer dimensions and that summarizes well the original data.

To illustrate, you will use the sample dataset MNIST Train 60k 28x28, which contains 60,000 examples of handwritten digits that have been size-normalized and centered in a fixed-size image 28x28 pixels; as a result there are 784 attributes (see Figure 3-25).

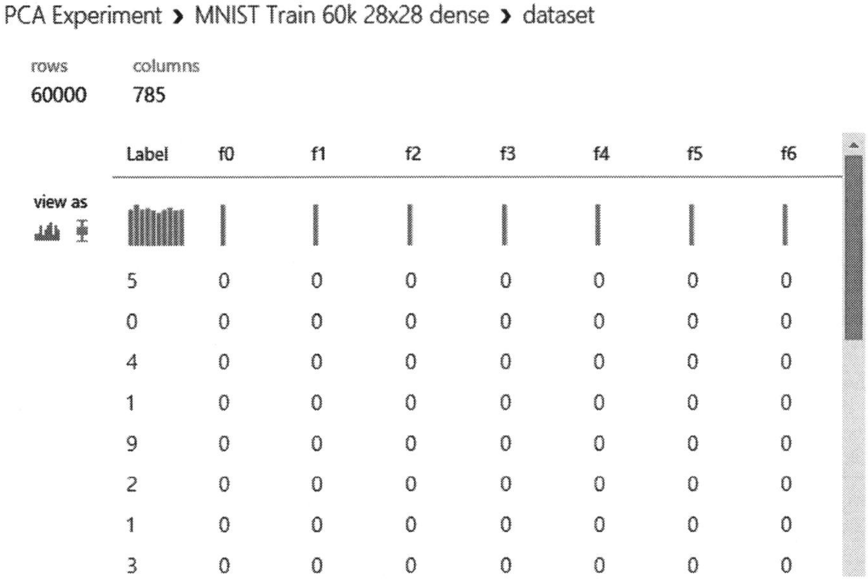

PCA Experiment › MNIST Train 60k 28x28 dense › dataset

rows 60000 columns 785

view as

Label	f0	f1	f2	f3	f4	f5	f6
5	0	0	0	0	0	0	0
0	0	0	0	0	0	0	0
4	0	0	0	0	0	0	0
1	0	0	0	0	0	0	0
9	0	0	0	0	0	0	0
2	0	0	0	0	0	0	0
1	0	0	0	0	0	0	0
3	0	0	0	0	0	0	0

Figure 3-25. *Display of MNIST dataset prior to PCA*

You will use the Principal Component Analysis module in Azure ML Studio, which is found in the Scale and Reduce drawer under the Data Transformation category (see Figure 3-26).

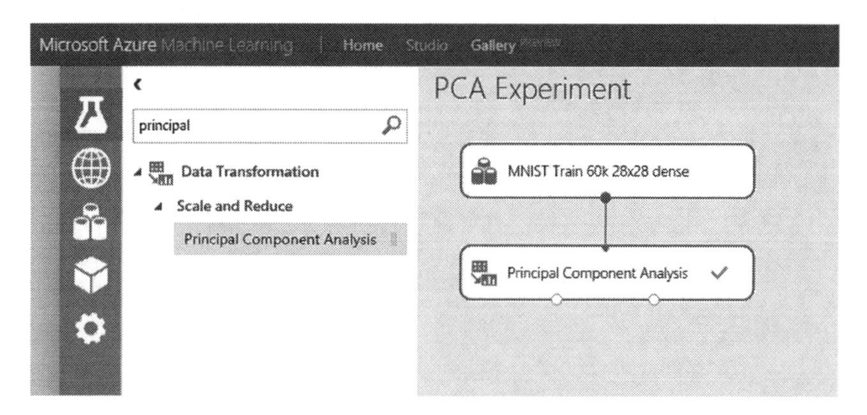

Figure 3-26. *The Principal Componet Analysis module*

To configure the Principal Component Analysis module, you simply need to connect the dataset, then select the columns on which the PCA analysis is to operate, and specify the number of dimensions the reduced dataset is to have. In selecting columns over which PCA is to operate, ensure that you do not select the class label and do not select features highly valuable for classification, perhaps identified through Pearson Correlation or identified through your own experiments. You should never use the class label to create a feature, nor should you lose highly valuable feature(s). Select the remaining undifferentiated features as input for PCA to operate on.

For illustration purposes, you will select 15 principal components to represent the datasets. Once complete, you can visualize these principal components, as illustrated in Figure 3-27, from the left output pin and then use this reduced dataset for machine learning.

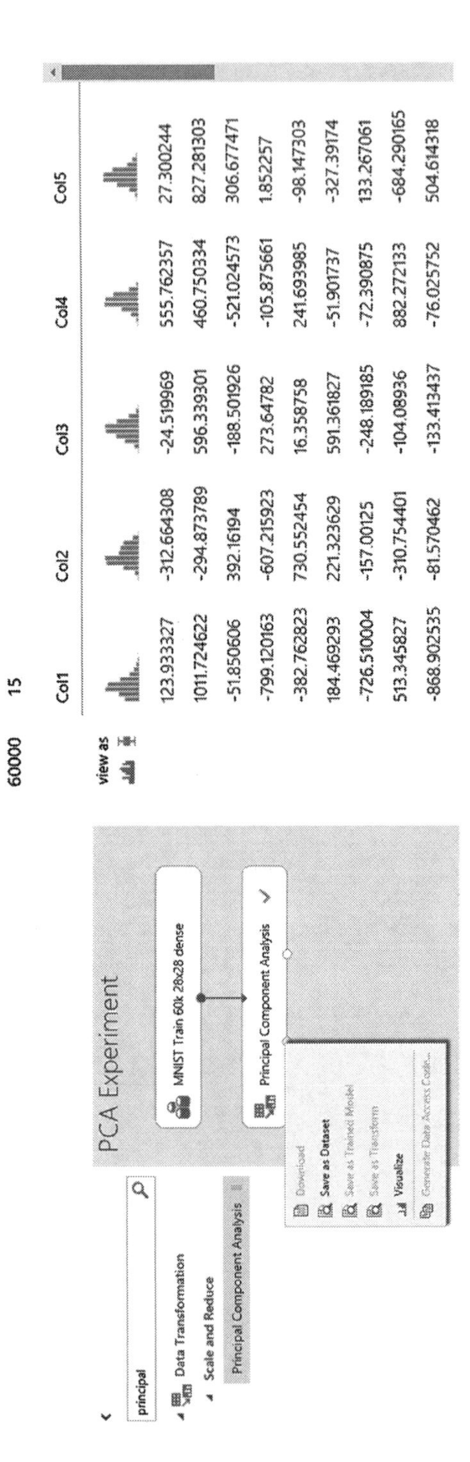

Col1	Col2	Col3	Col4	Col5
123.933327	-312.664308	-24.519969	555.762357	27.300244
1011.724622	-294.873789	596.339301	460.750334	827.281303
-51.850606	392.16194	-188.501926	-521.024573	306.677471
-799.120163	-607.215923	273.64782	-105.875661	1.852257
-382.762823	730.552454	16.358758	241.693985	-98.147303
184.469293	221.323629	591.361827	-51.901737	-327.39174
-726.510004	-157.00125	-248.189185	-72.390875	133.267061
513.345827	-310.754401	-104.08936	882.272133	-684.290165
-868.902535	-81.570462	-133.413437	-76.025752	504.614318

Figure 3-27. *Fifteen principal components of the dataset after PCA*

Summary

In this chapter, we described different methods by which input data can be processed using Azure ML to prepare for machine learning methods and improve the quality of the resulting model. We covered cleaning and processing your training data, including how to deal with missing data, class imbalance, and data normalization, followed by selecting and engineering features to increase the efficiency of the machine learning algorithm you have selected. Data preparation is a large subject that can involve a lot of iterations, experiments, and analysis. You will spend considerable time and effort in this phase of building a predictive model. But getting proficient in data preparation will make you a master at building predictive models. For now, consider the steps described in this chapter when preparing data and always be looking for ways to transform data and impart your insights on the data and your understanding of the domain so the learning algorithm can do its magic.

CHAPTER 4

Integration with R

This chapter will introduce R and show how it is integrated with Microsoft Azure Machine Learning. Through simple examples, you will learn how to write and run your own R code when working with Azure Machine Learning. You will also learn the R packages supported by Azure Machine Learning, and how you can use them in the Azure Machine Learning Studio (ML Studio).

R in a Nutshell

R is an open source statistical programming language that is commonly used by the computational statistics and data science community for solving an extensive spectrum of business problems. These problems span the following areas:

- Bioinformatics (think genome analysis)
- Actuarial sciences (such as figuring out risk exposures for insurance, finance, and other industries)
- Telecommunication (analyzing churn in corporate and consumer customer base, fraudulent SIM card usage, or mobile usage patterns)
- Finance and banking (such as identifying fraud in financial transactions)
- Manufacturing (predicting hardware component failure times), and many more

R empowers users with a toolbox of R packages that provides powerful capabilities for data analysis, visualization, and modeling. The Comprehensive R Archive Network (CRAN) is a large collection of more than 5,000 R packages. Besides CRAN, there are many other R packages available on Github (`https://github.com/trending?l=r`) and specialized R packages for bioinformatics in the Bioconductor R repository (`www.bioconductor.org/`).

Note R was created at the University of Auckland by George Ross Ihaka and Robert Gentleman in 1994. Since R's creation, many leading computer scientists and statisticians have fueled R's success by contributing to the R codebase or providing R packages that enable R users to leverage the latest techniques for statistical analysis, visualization, and data mining. This has propelled R to become one of the languages for data scientists. Learn more about R at `www.r-project.org/`.

Given the momentum in the data science community in using R to tackle machine learning problems, it is super important for a cloud-based machine learning platform to empower data scientists to continue using the familiar R scripts that they have written, and continue to be productive. Currently, more than 400 R packages are supported by Azure Machine Learning. Table 4-1 shows a subset of the R packages currently supported. These R packages enable you to model a wide spectrum of machine learning problems from market basket analysis, classification, regression, forecasting, and visualization.

Table 4-1. *R Packages Supported by Azure Machine Learning*

R Packages	Description
Arules	Frequent itemsets and association rule mining
FactoMineR	Data exploration
Forecast	Univariate time series forecasts (exponential smoothing and automatic ARIMA)
ggplot2	Graphics
Glmnet	Linear regression, logistics and multinomial regression models, poisson regression, and Cox Model
Party	Tools for decision trees
randomForest	Classification and regression models based on a forest of trees
Rsonlp	General non-linear optimization using augmented Lagrange multipliers
Xts	Time series
Zoo	Time series

■ **Tip** To get the complete list of installed packages, create a new experiment in Cloud Machine Learning, use the **Execute R Script** module, provide the following script in the body of the **Execute R Script** module, and run the experiment. After the experiment

```
out <- data.frame(installed.packages())
maml.mapOutputPort("out")
```

completes, right-click the left output portal of the module and select **Visualize**. The packages that have been installed in Cloud Machine Learning will be listed.

Azure Machine Learning provides two R language modules to enable you to integrate R into your machine learning experiments: the Execute R Script and Create R Model modules.

The Execute R Script module enables you to specify input datasets (at most two datasets), an R script, and a ZIP file containing a set of R scripts (optional). After the module processes the data, it produces a result dataset and an R device output. In Azure Machine Learning, the R scripts are executed using R 3.1.0.

The Create R Model module can be used to create an untrained model using an R script that you provide. Your R code can only contain packages that release in Azure Machine Learning. The beauty of this module is that you can use the Train Model module to train it just as you do with any other learning algorithm. You can also use the Score Model module to test your model.

■ **Note** The R Device output shows the console output and graphics that are produced during the execution of the R script. For example, in the R script, you may have used the R `plot()` function. The output of `plot()` can be visualized when you right-click the R Device output and choose **Visualize**.

Azure ML Studio enables you to monitor and troubleshoot the progress of the experiment. Once the execution has completed, you can view the output log of each run of the R module. The output log will also enable you to troubleshoot issues if the execution failed.

In this chapter, you will learn how to integrate R with Azure Machine Learning. Through the use of simple examples and datasets available in ML Studio, you will gain essential skills to unleash the power of R to create exciting and useful experiments with Azure Machine Learning. Let's get started!

Building and Deploying Your First R Script

To build and deploy your first R script module, first you need to create a new experiment. After you create the experiment, you will see the **Execute R Script** module that is provided in the Azure Machine Learning Studio (ML Studio). The script will be executed by Azure Machine Learning using R 3.1.0 (the version that was installed on Azure Machine Learning at the time of this book). Figure 4-1 shows the R Language modules.

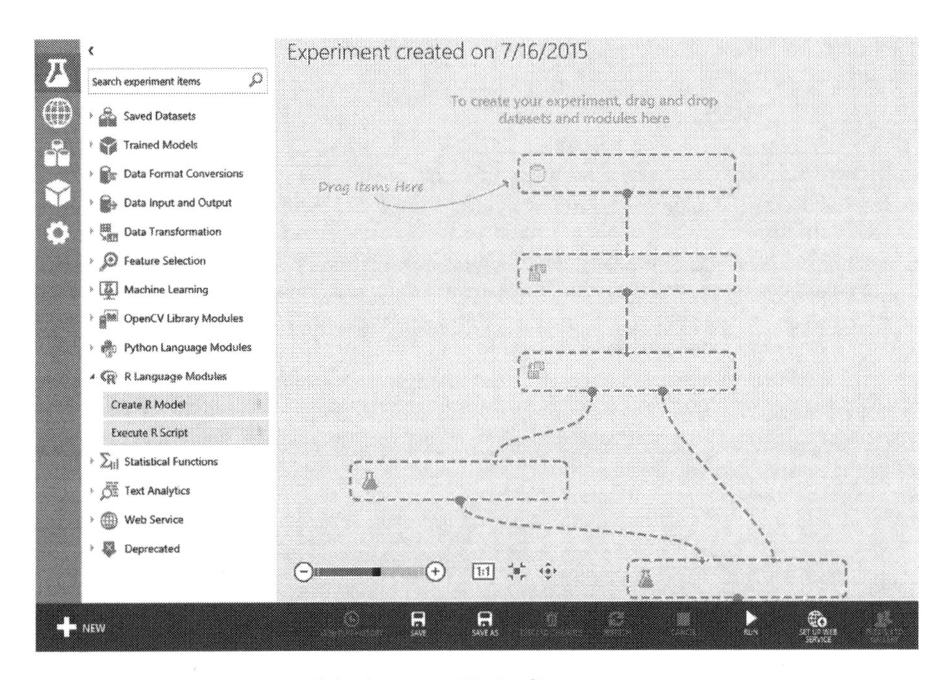

Figure 4-1. *R Language modules in Azure ML Studio*

In this section, you will learn how to use the **Execute R Script** module to perform sampling on a dataset. Follow these steps.

1. From the toolbox, expand the **Saved Datasets** node, and click the **Adult Census Income Binary Classification** dataset. Drag and drop it onto the experiment design area.

2. From the toolbox, expand the **R Language Modules** node, and click the **Execute R Script** module. Drag and drop it onto the experiment design area.

3. Connect the dataset to the **Execute R Script** module. Figure 4-2 shows the experiment design.

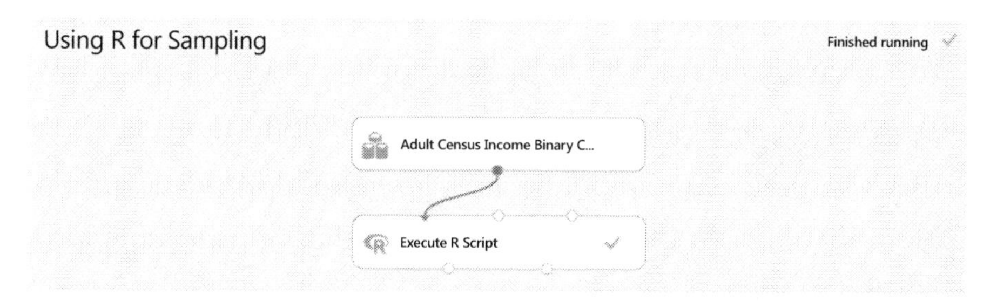

Figure 4-2. *Using the Execute R Script module to perform sampling on the Adult Census Income Binary Classification dataset*

The **Execute R Script** module provides two input ports for datasets that can be used by the R script. In addition, it allows you to specify a ZIP file that contains the R source script files that are used by the module. You can edit the R source script files on your local machine and test them. Then, you can compress the needed files into a ZIP file and upload it to Azure Machine Learning through New ➤ Dataset ➤ From Local File path. After the **Execute R Script** module processes the data, the module provides two output ports: Result Dataset and an R Device. The Result Dataset corresponds to output from the R script that can be passed to the next module. The R Device output port provides you with an easy way to see the console output and graphics that are produced by the R interpreter.

Let's continue creating your first R script using ML Studio.

4. Click the **Execute R Script** module.
5. On the Properties pane, write the following R script to perform sampling:

```
# Map 1-based optional input ports to variables
dataset1 <- maml.mapInputPort(1) # class: data.frame
mysample <- dataset1[sample(1:nrow(dataset1), 50,
replace=FALSE),]

data.set = mysample
print (mysample)

# Select data.frame to be sent to the output Dataset port
maml.mapOutputPort("data.set");

R Script to perform sampling
```

To use the **Execute R Script** module, the following pattern is often used:

- Map the input ports to R variables or data frame.
- Main body of the R Script.
- Map the results to the output ports (see Figure 4-3).

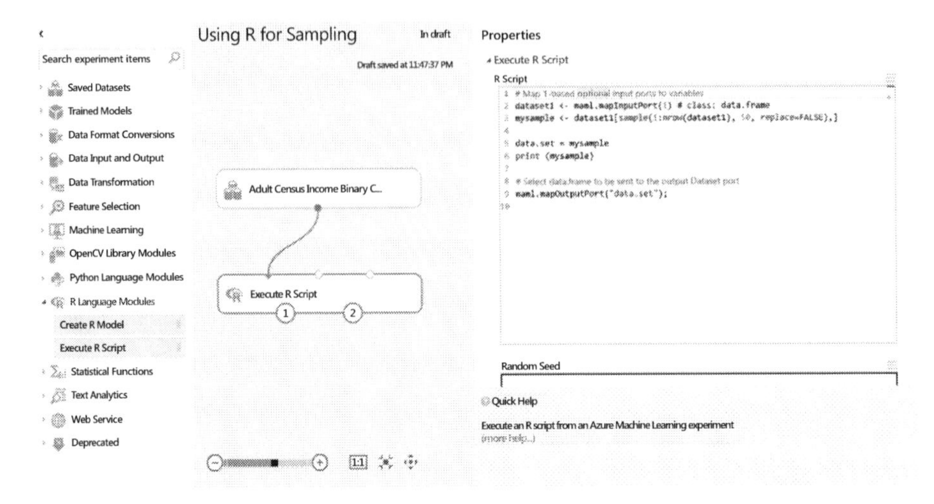

Figure 4-3. *Successful execution of the sampling experiment*

To see this pattern in action, observe that in the R script provided, the `maml.mapInputPort(1)` method is used to map the dataset that was passed in from the first input port of the module to an R data frame. Next, see the R script that is used to perform the sampling of the data. For debugging purposes, we also printed out the results of the sample. In the last step of the R script, the results are assigned to `data.set` and mapped to the output port using `maml.mapOutputPort("data.set")`.

You are now ready to run the experiment. To do this, click the Run icon at the bottom pane of Azure ML Studio. Figure 4-3 shows that the experiment has successfully completed execution.

Once the experiment has finished running, you can see the output of the R script. To do this, right-click the **Result Dataset**. Figure 4-4 shows the options available when you right-click **Result Dataset**. Choose **Visualize**.

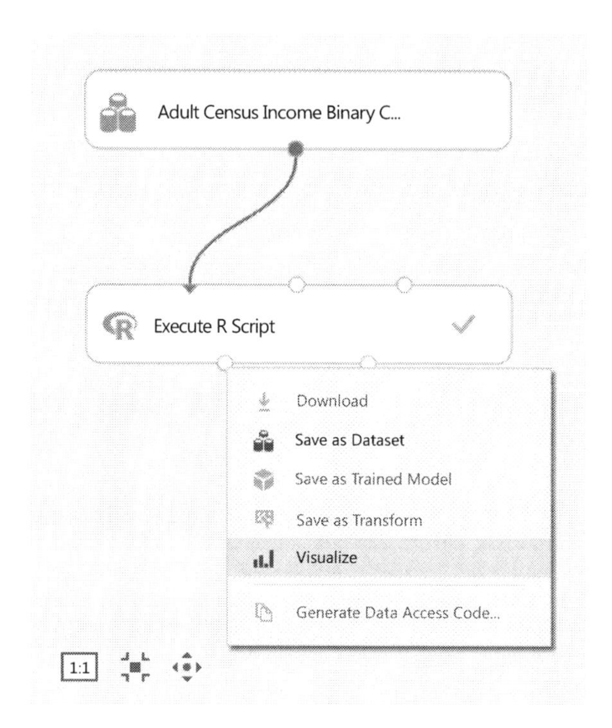

Figure 4-4. *Visualizing the output of the R script*

After you click **Visualize**, you will see that the sample consists of 50 rows. Each of the rows has 15 columns and the data distribution for the data. Figure 4-5 shows the visualization for the Result Dataset.

Ch04 - Sampling Example › Execute R Script › Result Dataset

rows 50 columns 15

age	workclass	fnlwgt	education	education-num	marital-status
34	Private	48014	Bachelors	13	Separated
18		216508	HS-grad	9	Never-married
19	Private	184737	HS-grad	9	Never-married
59	Private	182460	Bachelors	13	Married-civ-spouse
37	Private	389725	12th	8	Divorced

Statistics

Mean	40.08
Median	36.5
Min	17
Max	90
Standard Deviation	17.8564
Unique Values	36
Missing Values	0
Feature Type	Numeric Feature

Visualizations

age
Histogram
compare to None

Figure 4-5. *Visualization of Result Dataset produced by the R script*

Congratulations, you have just successfully completed the integration of your first simple R script in Azure Machine Learning! In the next section, you will learn how to use Azure Machine Learning and R to create a machine learning model.

■ **Note** Learn more about R and Machine Learning at `http://ocw.mit.edu/courses/sloan-school-of-management/15-097-prediction-machine-learning-and-statistics-spring-2012/lecture-notes/MIT15_097S12_lec02.pdf`.

Using R for Data Preprocessing

In many machine learning tasks, dimensionality reduction is an important step that is used to reduce the number of features for the machine learning algorithm. Principal component analysis (PCA) is a commonly used dimensionality reduction technique. PCA reduces the dimensionality of the dataset by finding a new set of variables (principal components) that are linear combinations of the original dataset, and are uncorrelated with all other variables. In this section, you will learn how to use R for preprocessing the data and reducing the dimensionality of dataset.

Let's get started with using the **Execute R Script** module to perform principal component analysis of one of the sample datasets available in Azure ML Studio. Follow these steps.

1. Create a new experiment.
2. From Saved Datasets, choose **CRM Dataset Shared** (see Figure 4-6).

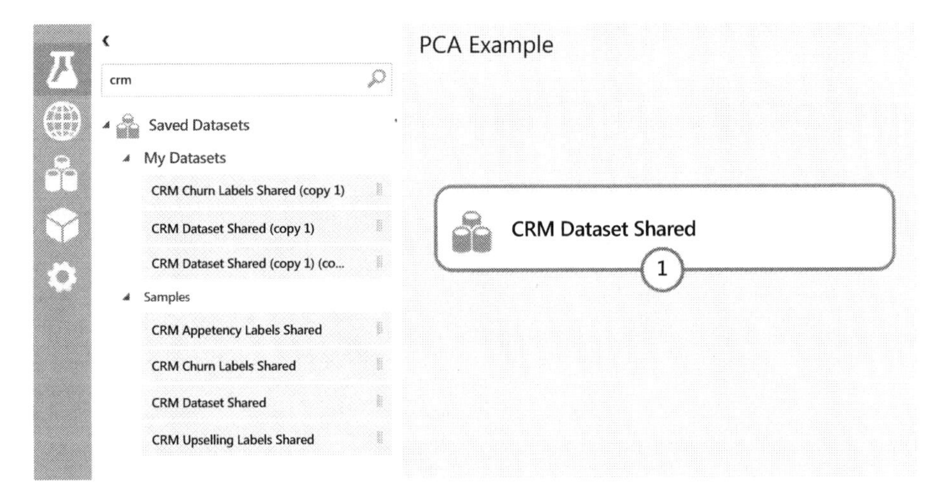

Figure 4-6. *Using the sample dataset, CRM Dataset Shared*

3. Right-click the output node, and choose **Visualize**. From the visualization shown in Figure 4-7, you will see that are 230 columns.

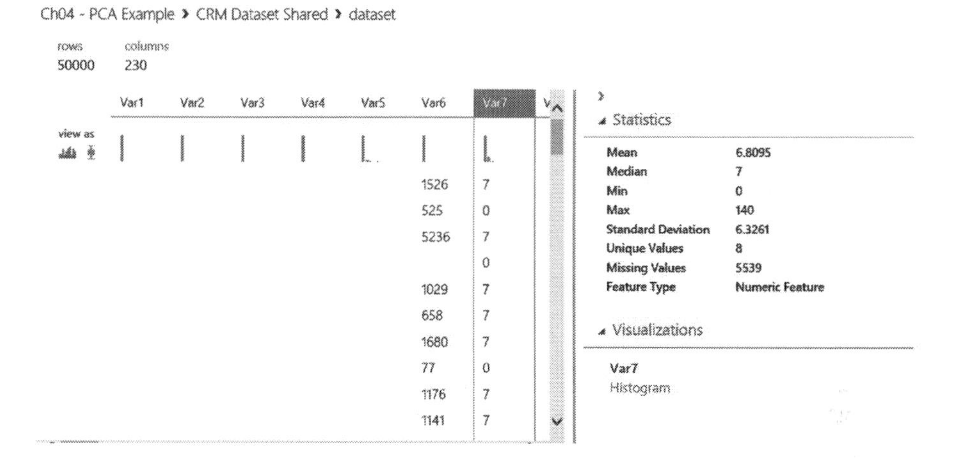

***Figure 4-7.** Initial 230 columns from the CRM Dataset Shared*

4. Before you perform PCA, you need to make sure that the inputs to the **Execute R Script** module are numeric. For the sample dataset of CRM Dataset Shared, you know that the first 190 columns are numeric and the remaining 40 columns are categorical. You first use the **Project Columns**, and set the **Selected Column** | column indices to be 1-190.

5. Next, use the **Missing Value Scrubber** module to replace the missing value with 0, as follows:

    a. Cleaning Mode: Custom substitution value

    b. Replacement Value: 0

6. Once the missing values have been scrubbed, use the **Metadata Editor** module to change the data type for all the columns to be Integer.

7. Next, drag and drop the **Execute R Script** module to the design surface and connect the modules, as shown in Figure 4-8.

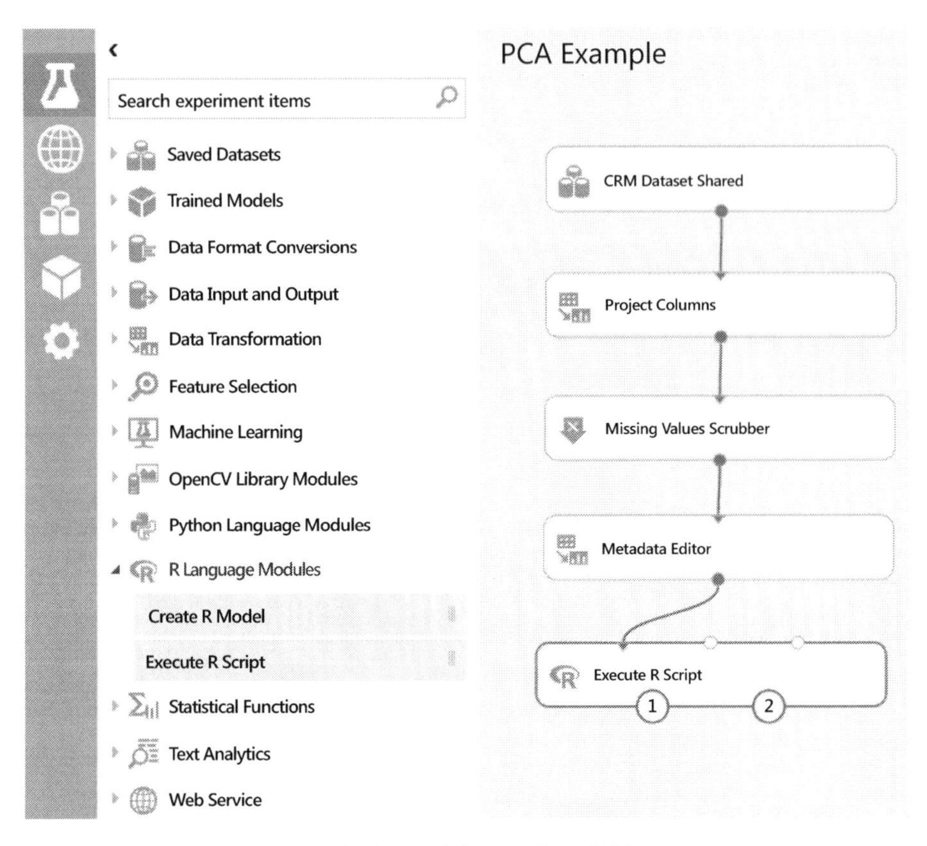

Figure 4-8. *Using the Execute R Script module to perform PCA*

8. Click the **Execute R Script** module. You can provide the following script that will be used to perform PCA:

```
# Map 1-based optional input ports to variables
dataset1 <- maml.mapInputPort(1)
# Perform PCA on the first 190 columns
pca = prcomp(dataset1[,1:190])

# Return the top 10 principal components
top_pca_scores = data.frame(pca$x[,1:10])
data.set = top_pca_scores
plot(data.set)
# Select data.frame to be sent to the output Dataset port
maml.mapOutputPort("data.set");
```

Figure 4-9 shows how to use the script in Azure ML Studio.

Properties

◢ Execute R Script

R Script

```
1  # Map 1-based optional input ports to variables
2  dataset1 <- maml.mapInputPort(1) # class: data.frame
3
4  # Perform PCA on the first 190 columns
5  pca = prcomp(dataset1[,1:190])
6
7  # Return the top 10 principal components
8  top_pca_scores = data.frame(pca$x[,1:10])
9  data.set = top_pca_scores
10
11 # You'll see this output in the R Device port.
12 # It'll have your stdout, stderr and PNG graphics device(s).
13 plot(data.set);
14
15 # Select data.frame to be sent to the output Dataset port
16 maml.mapOutputPort("data.set");
```

Quick Help

Execute the given R script using R 3.1.0
(more help...)

Figure 4-9. *Performing PCA on the first 190 columns*

9. Click **Run**.

10. Once the experiment has successfully executed, click the **Result dataset** node of the **Execute R Script** module. Click **Visualize**.

11. Figure 4-10 shows the top ten principal components for the CRM Shared dataset. The principal components are used in the subsequent steps for building the classification model.

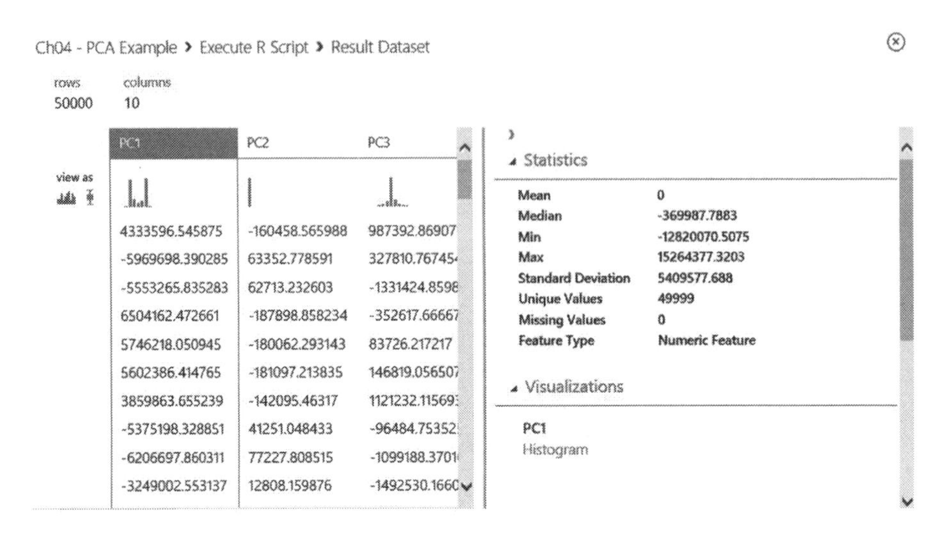

Figure 4-10. *Principal components identified for the CRM Shared dataset*

■ **Tip** You can also provide a script bundle containing the R script in a ZIP file, and use it in the **Execute R Script** module.

Using a Script Bundle (ZIP)

If you have an R script that you have been using and want to use it as part of the experiment, you can ZIP up the R script, and upload it to Azure ML Studio as a dataset. To use a script bundle with the **Execute R Script** module, you will first need to package up the file in ZIP format, so follow these steps.

1. Navigate to the folder containing the R scripts that you intend to use in your experiment (see Figure 4-11).

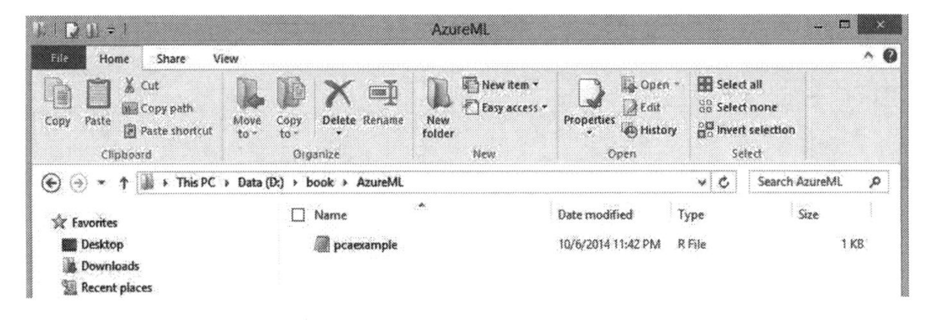

Figure 4-11. *Folder containing the R script pcaexample.r*

2. Select all the files that you want to package up and right-click. In this example, right-click the file **pcaexample.r**, and choose **Send to Compressed (zipped) folder**, as shown in Figure 4-12.

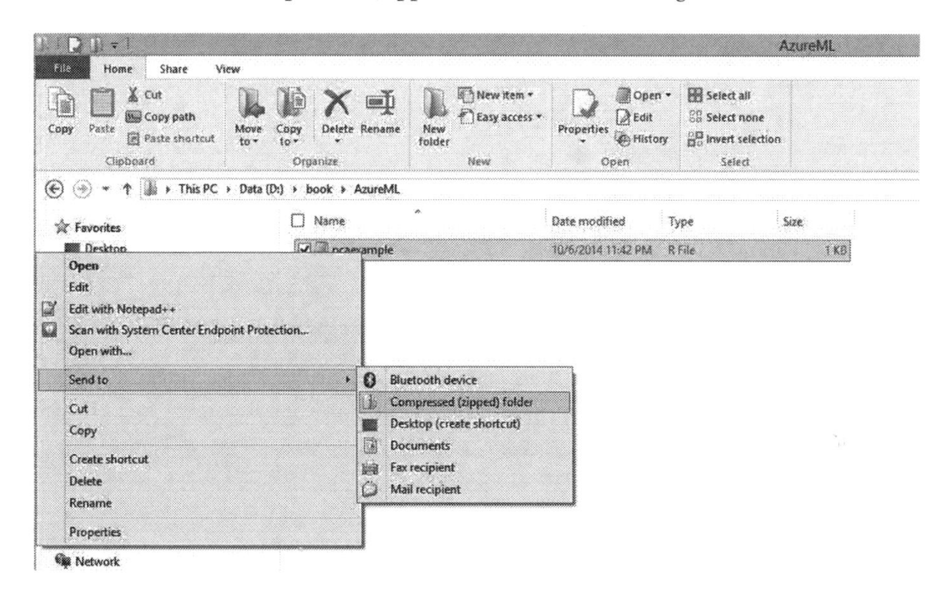

Figure 4-12. *Packaging the R scripts as a ZIP file*

3. Next, upload the ZIP file to Azure ML Studio. To do this, choose **New DATASET ➤ From Local File**. Select the ZIP file that you want to upload, as shown in Figures 4-13 and 4-14.

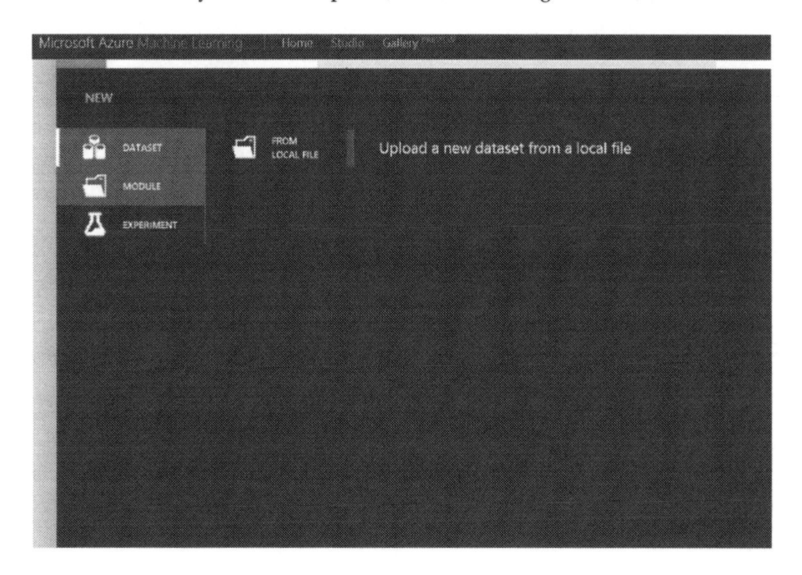

Figure 4-13. *New Dataset ➤ From Local File*

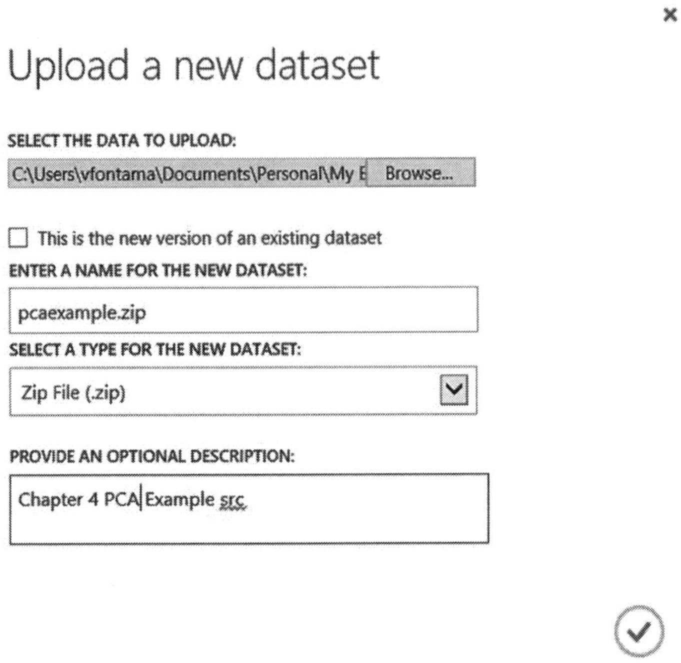

×

Upload a new dataset

SELECT THE DATA TO UPLOAD:

C:\Users\vfontama\Documents\Personal\My E Browse...

☐ This is the new version of an existing dataset

ENTER A NAME FOR THE NEW DATASET:

pcaexample.zip

SELECT A TYPE FOR THE NEW DATASET:

Zip File (.zip)

PROVIDE AN OPTIONAL DESCRIPTION:

Chapter 4 PCA Example src

Figure 4-14. *Uploading a new dataset*

4. Once the dataset has been uploaded, you can use it in your experiment. To do this, from **Saved Datasets**, select the **uploaded ZIP file**, and drag and drop it to the experiment.

5. Connect the uploaded ZIP file to the **Execute R Script** module – **Script Bundle (ZIP)** input, as shown in Figure 4-15.

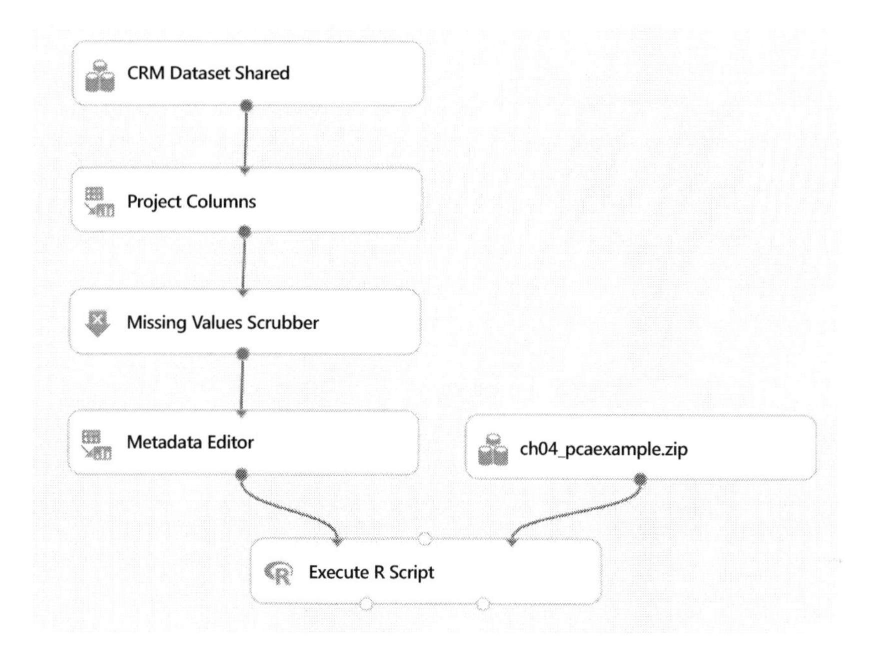

Figure 4-15. *Using the script bundle as an input to the Execute R Script module*

6. In the **Execute R Script** module, specify where the R script can be found, as follows and as shown in Figure 4-16:

```
# Map 1-based optional input ports to variables
dataset1 <- maml.mapInputPort(1)
# Contents of optional ZIP port are in ./src/
source("src/pcaexample.r");

# Select data.frame to be sent to the output Dataset port
maml.mapOutputPort("data.set");
```

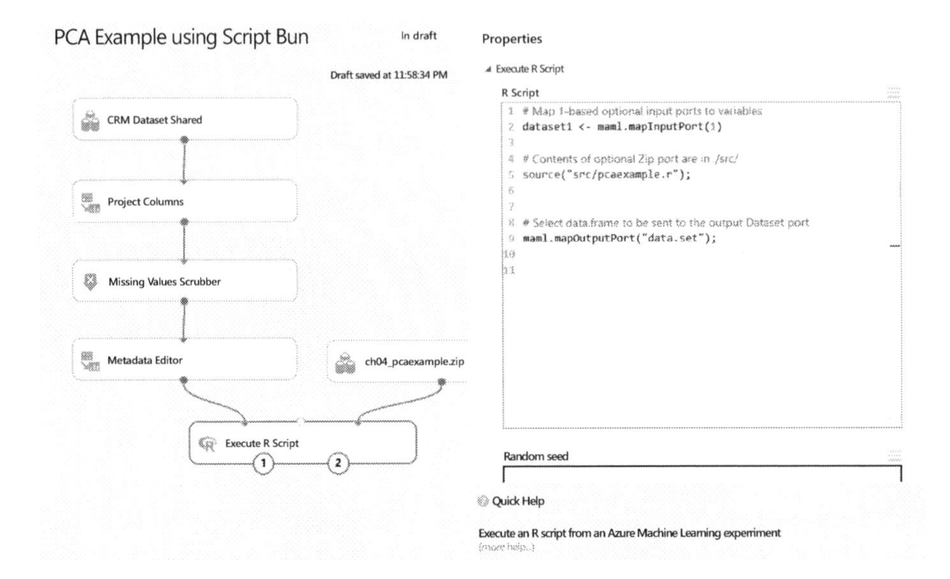

Figure 4-16. *Using the script bundle as inputs to the Execute R Script module*

7. You are now ready to run the experiment using the R script in the ZIP file that you have uploaded to ML Studio.

   Using the script bundle allows you to easily reference an R script file that you can test outside of Azure ML Studio. In order to update the script, you will have to re-upload the ZIP file.

Building and Deploying a Decision Tree Using R

In this section, you will learn how to use R to build a machine learning model. When using R, you can tap into the large collection of R packages that implement various machine learning algorithms for classification, clustering, regression, K-Nearest Neighbor, market basket analysis, and much more.

▒ **Note** When you use the machine learning algorithms available in R and execute them using the **Execute R** module, you can only visualize the model and parameters. You cannot save the trained model and use it as input to other Azure ML Studio modules.

Using Azure ML studio, several R machine learning algorithms are provided. For example, you can use the **auto.arima function** from the Forecast package to build an optimal autoregressive moving average model for a univariate time series. You can also use the **knn function** to create a k-nearest neighbor (KNN) classification model.

In this section, you will learn how to use an R package called **rpart** to build a decision tree. The **rpart** package provides you with recursive partitioning algorithms for performing classification and regression.

■ **Note** Learn more about the rpart R package at `http://cran.r-project.org/web/packages/rpart/rpart.pdf`.

For this exercise, you will use the Adult Census Income Binary Classification dataset. Let's get started.

1. From the toolbox, drag and drop the following modules on the experiment design area:

   a. **Adult Census Income Binary Classification** dataset (available under **Saved Datasets**)

   b. **Project Columns** (available from Data **Transformation ➤ Manipulation**)

   c. **Execute R Script** module (found under **R Language** modules)

2. Connect the **Adult Census Income Binary Classification** dataset to **Project Columns**.

3. Click **Project Columns** to select the columns that will be used in the experiment. Select the following columns: age, sex, education, income, marital-status, occupation.

Figure 4-17 shows the selection of the columns and Figure 4-18 shows the completed experiment and the `Project Columns` properties.

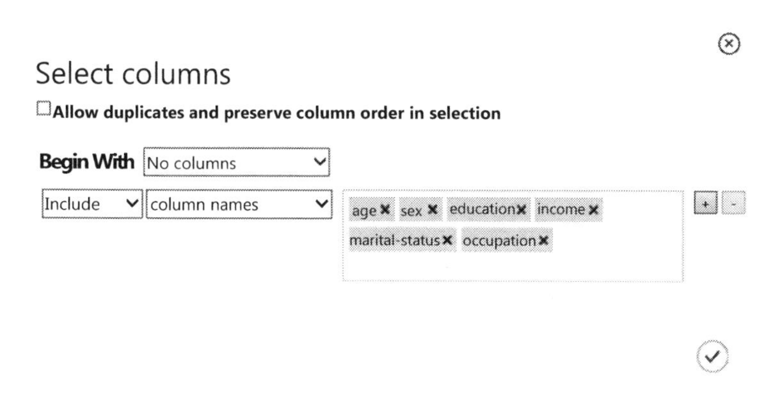

***Figure 4-17.** Selecting columns that will be used*

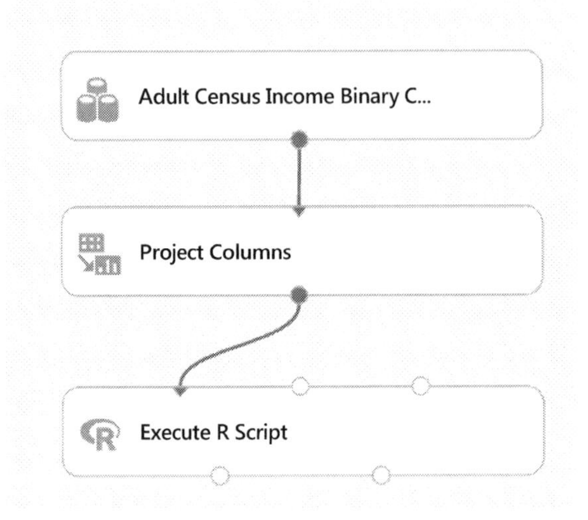

Figure 4-18. *Complete Experiment*

4. Connect the **Project Columns** to **Execute R Script**.

5. Click **Execute R Script** and provide the following script:

```
library(rpart)

# Map 1-based optional input ports to variables
Dataset1 <- maml.mapInputPort(1) # class: data.frame

fit <- rpart(income ~ age + sex + education + occupation ,
method="class", data=Dataset1)

# display the results, and summary of the splits
printcp(fit)
plotcp(fit)
summary(fit)

# plot the decision tree
plot(fit, uniform=TRUE, margin = 0.1,compress = TRUE,
main="Classification Tree for Census Dataset")
text(fit, use.n=TRUE, all=TRUE, cex=0.8, pretty=1)
```

```
data.set = Dataset1

# Select data.frame to be sent to the output Dataset port
maml.mapOutputPort("data.set");
```

In the R script, first load the rpart library using `library()`. Next, map the dataset that is passed to the **Execute R Script** module to a data frame.

To build the decision tree, you will use the `rpart` function. Several types of methods are supported by rpart: `class` (classification), and anova (regression). In this exercise, you will use `rpart` to perform classification (such as `method="class"`), as follows:

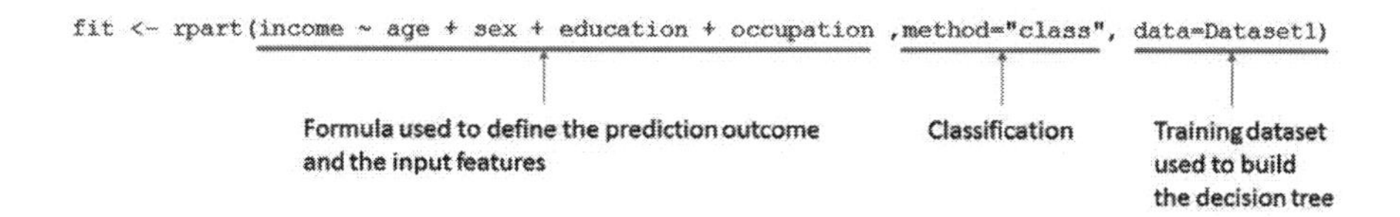

The formula specified uses the following format: predictionVariable ~ inputfeature1 + inputfeature2 + ...

After the decision tree has been constructed, the R script invokes the `printcp()`, `plotcp()`, and `summary()` functions to display the results and a summary of each of the split values in the tree. In the R script, the `plot()` function is used to plot the rpart model. By default, when the rpart model is plotted, an abbreviated representation is used to denote the split value. In the R script, the setting of `pretty=1` is added to enable the actual split values to be shown (see Figure 4-19).

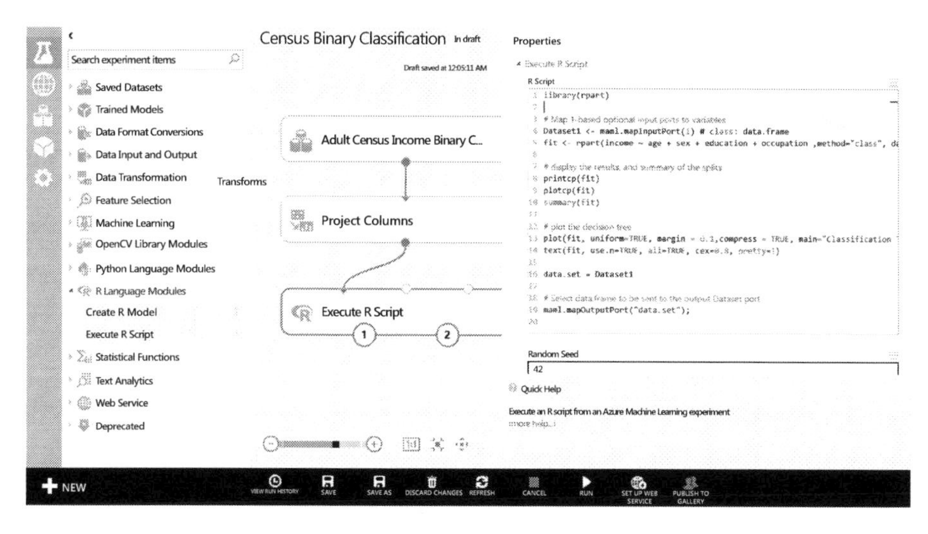

Figure 4-19. *The R script in Azure ML Studio*

You are now ready to run the experiment. To do this, click the Run icon at the bottom pane of Azure ML Studio. Once the experiment has executed successfully, you can view the details of the decision tree and also visualize the overall shape of the decision tree.

To view the details of the decision tree, click the **Execute R Script** module and **View output log** (shown in Figure 4-20) in the **Properties** pane.

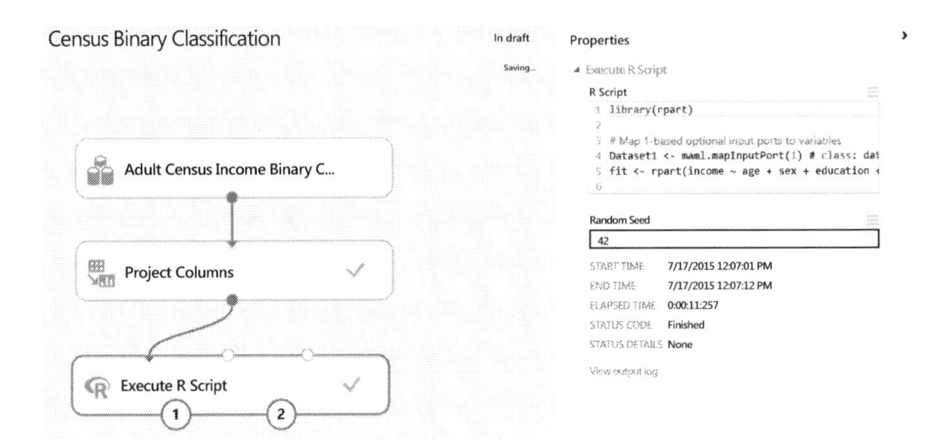

Figure 4-20. *View output log for Execute R Script*

A sample of the output log is shown:

```
[ModuleOutput] Classification tree:
[ModuleOutput]
[ModuleOutput]
[ModuleOutput]
[ModuleOutput] Variables actually used in tree construction:
[ModuleOutput]
[ModuleOutput] [1] age       education  occupation sex
[ModuleOutput]
[ModuleOutput]
...
[ModuleOutput]   Primary splits:
[ModuleOutput]
[ModuleOutput]       education  splits as  LLLLLLLLLRRLRLRL,
improve=1274.3680, (0 missing)
[ModuleOutput]
[ModuleOutput]       occupation splits as  LLLRLLLLLRRLLL,
improve=1065.9400, (1843 missing)
[ModuleOutput]
[ModuleOutput]       age        < 29.5 to the left,  improve= 980.1513,
(0 missing)
[ModuleOutput]
[ModuleOutput]       sex        splits as  LR, improve= 555.3667,
(0 missing)
```

To visualize the decision tree, click the **R Device** output port (the second output of the **Execute R Script** module), and you will see the decision tree that you just constructed using `rpart` (shown in Figure 4-21).

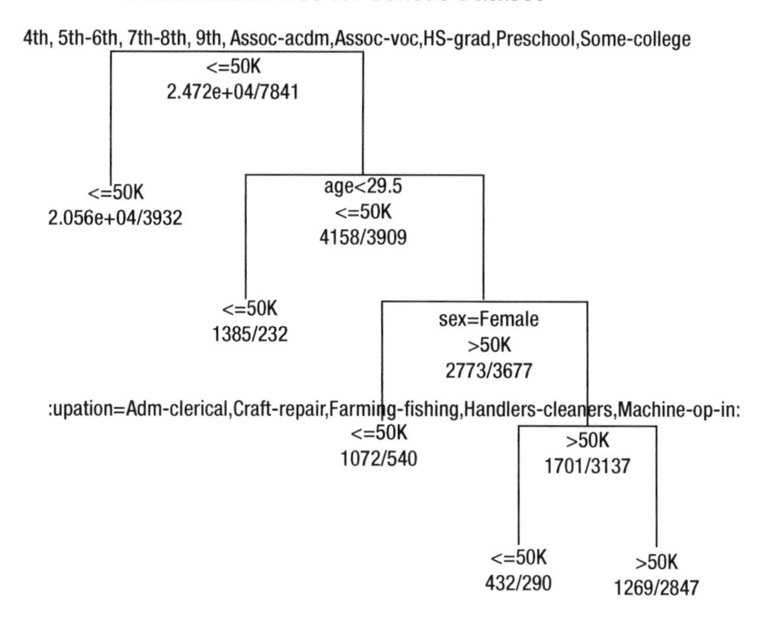

Figure 4-21. *Decision tree constructed using rpart*

■ **Note** Azure Machine Learning provides a good collection of saved datasets that can be used in your experiments. In addition, you can also find an extensive collection of datasets at the UCI Machine Learning Repository at `http://archive.ics.uci.edu/ml/datasets.html`.

Summary

In this chapter, you learned about the exciting possibilities offered by R integration with Azure Machine Learning. You learned how to use the different R Language modules in Azure ML Studio. As you designed your experiment using R, you learned how to map the inputs and outputs of the module to R variables and data frames. Next, you learned how to build your first R script to perform data sampling and how to visualize the results using the built-in data visualization tools available in Azure ML Studio. With that as foundation, you moved on to building and deploying experiments that use R for data preprocessing through an R Script bundle and building decision trees.

CHAPTER 5

Integration with Python

This chapter shows you how to use Python in Azure Machine Learning (Azure ML). Using simple examples, you will learn how to integrate Python as part of an Azure ML experiment. This enables you to tap into the powerful capabilities offered by various Python libraries, such as NumPy, SciPy, pandas, scikit-learn, and many more, directly in an Azure ML experiment.

Overview

Python is an elegant and powerful interpreted programming language. Both R and Python rank amongst the top programming language of choice for data scientists and developers. Python empowers data scientists and developers to use it in early experimentation and to have confidence in deploying it into production environments. Many large organizations, including YouTube, Industrial Light & Magic, IronPort, HomeGain, and Eve Online, have used Python successfully in production (`www.python.org/about/quotes/`).

Since the invention of Python by Guido van Rossum at Centrum Wiskunde & Informatica (CWI) in the late 1980s, many powerful Python libraries for performing data analysis have been made available, including the following:

- **NumPy** (Numerical Python): Provides fast and efficient multidimensional array operations, a rich set of linear algebra methods, and random number generators.
- **pandas**: Provides powerful data processing capabilities (reshaping, slicing and dicing, aggregations, filtering) for structured data sets.
- **SciPy**: Provides a rich set of mathematical and utility functions. SciPy is organized as subpackages, and each subpackage is used for data analysis. The capabilities offered by the subpackages range from fast Fourier transforms and interpolation methods to a large collection of statistical distributions and functions.

- **scikit-learn**: Provides an extensive set of machine learning capabilities including classification, regression, cluster, dimensionality reduction, model selection, and data preprocessing.
- **matplotlib**: Enables users to easily generate charts and visualizations.

Besides these general-purpose Python libraries, there are many domain-specific Python libraries available for applications ranging from computational biology and bioinformatics to astronomy. For example, BioPython is used by data scientists in the bioinformatics community for processing biological sequence files as well as working and visualizing phylogenetic trees. As of May 2015, the Python Package Index (`https://pypi.python.org/pypi`), a repository for Python libraries, contained 59,717 Python packages.

If you are a Python expert, this chapter enables you to quickly understand how you can leverage familiar tools and integrate the Python scripts into an Azure ML experiment. If you are new to Python, this chapter helps you gain the basics of using Python for data analysis and pre-processing, and sets you up for a successful Python learning journey. This chapter is not meant to be a replacement of the many excellent Python online resources and books; in fact, this chapter offers references to useful Python online resources that you can tap into.

Let's get started with understanding how Python can be used with Azure ML.

Python Jumpstart

Python provides an interactive command-line shell that enables you to work with Python code. In addition, the IPython Notebook provides a rich, interactive browser-based environment for executing Python code and visualizing the results. In this chapter, you will run the Python code in an IPython notebook.

To get started, download Anaconda from `http://continuum.io/downloads`. Anaconda provides a free Python distribution that you can use. In this chapter, we will focus on running IPython on Windows. In addition, the Anaconda installation installs the IPython notebook. After you have installed Anaconda, you can perform the following steps to create the IPython notebook.

1. Press the **Windows** button, and type **IPython**. Figure 5-1 shows the search results.

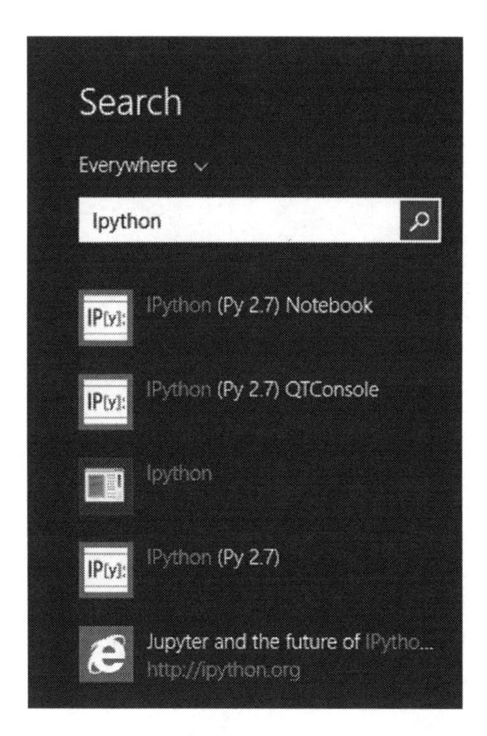

Figure 5-1. Searching for IPython Notebook

2. Start the IPython server by clicking the **IPython (Py 2.7) Notebook**.

3. Once the IPython server is started, it automatically navigates to `http://localhost:8888/tree`.

4. Once you are on the IPython home page, you can create a new Python notebook by clicking **New**, and choosing **Python 2 notebook**. Figure 5-2 shows how to create the notebook.

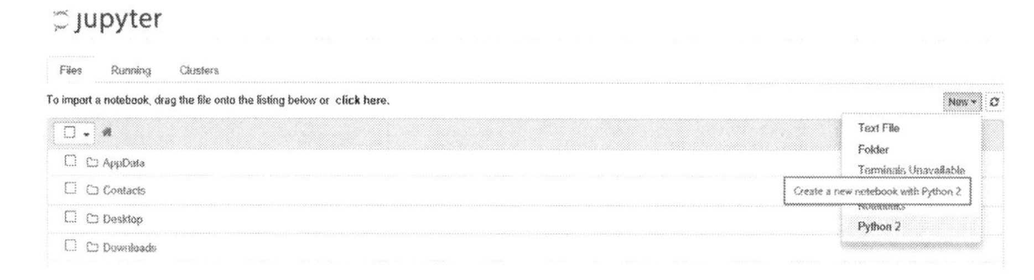

Figure 5-2. Creating a Python 2 notebook

5. After the notebook has been created, you can navigate to the new notebook.

6. Type `print "Hello World"` and click the **Run** button. Figure 5-3 shows the Hello World example.

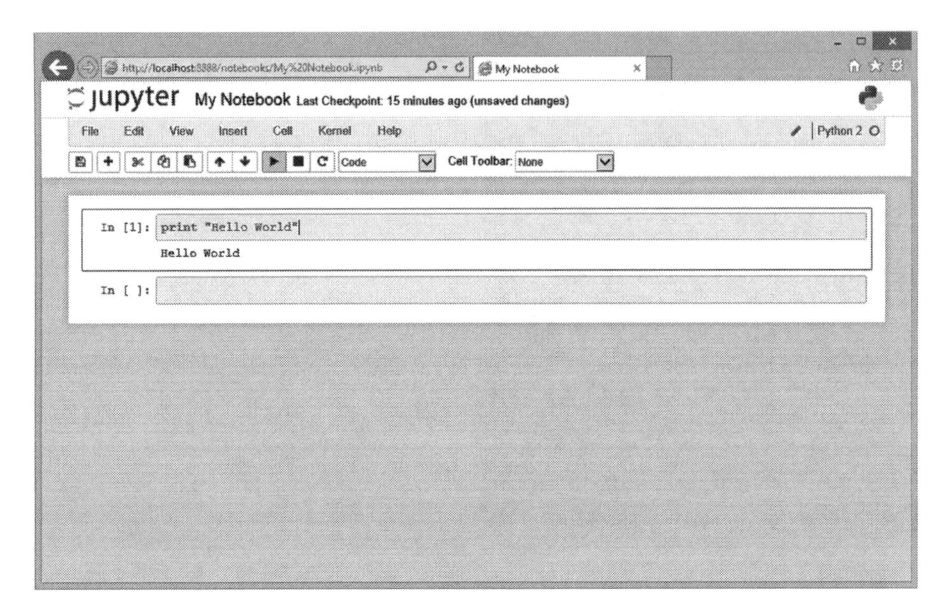

Figure 5-3. *Hello World Example*

You can use IPython notebook to develop your Python script and test it before integrating it with Azure ML. To use Azure ML with your Python environment, you will need to install the AzureML Python module. To do this, follow these steps.

1. Press the **Windows** button, and type **Command Prompt**. Figure 5-4 shows the search results.

Figure 5-4. *Using the command prompt*

2. Select **Command Prompt**, and navigate to the folder where you have installed Python. For this exercise, we will assume that Python is installed in `C:\Python34`.

   However, depending on your installation, you might find Python installed in the default installation folder, `C:\Users\localuser\AppData\Local\Continuum\Anaconda`.

3. Set the path using the command prompt:

   ```
   path=%path%;c:\python34\scripts
   ```

4. After you have set the path successfully, you can use the `pip` command to install the AzureML module, as follows:
   `pip install azureml`

After you have installed the azureml module successfully, you can use it in an IPython notebook. Figure 5-5 shows the azureml module installed successfully.

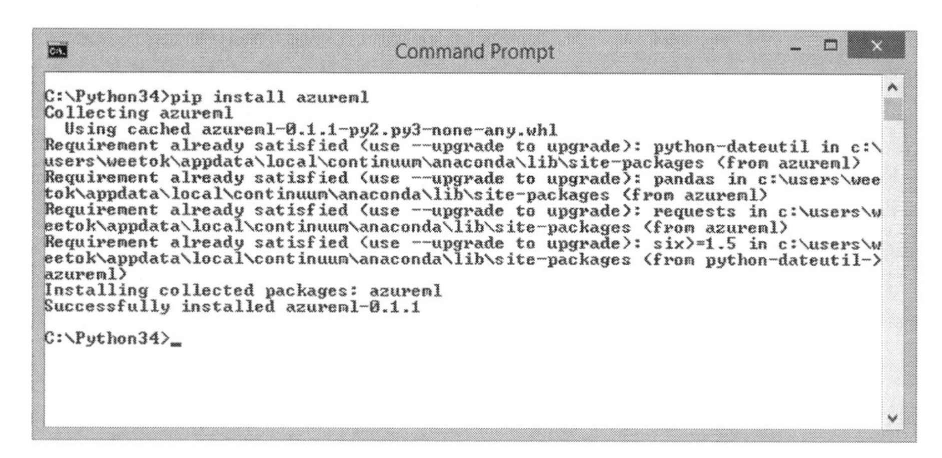

Figure 5-5. *Installing the azureml module*

In the next section, you will learn how to use the IPython notebook to access the data used in an Azure ML experiment.

▒ **Note** Learn more about Python at `www.python.org/` and design philosophies in the "Zen of Python" at `www.python.org/dev/peps/pep-0020/`.

Using Python in Azure ML Experiments

Let's get started on using IPython notebook to work with the experiments in ML Studio. But first, let's use one of the sample experiments from the gallery.

1. In your web browser, navigate to the following URL: `http://gallery.azureml.net/browse/?tags=["demand estimation"]`.

   You can also search for the experiment in the gallery by using the keywords **demand estimation**.

Figure 5-6 shows the sample experiment for demand estimation. This experiment shows how to build a regression model for bike rental demand.

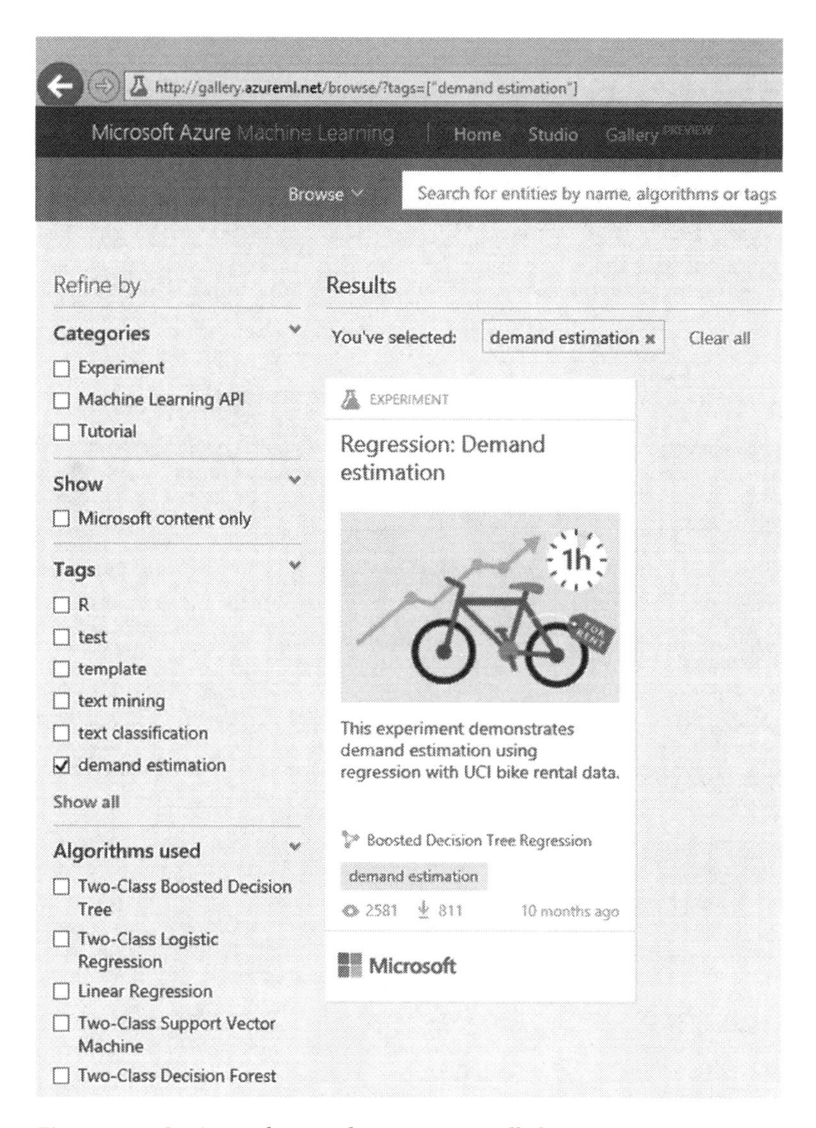

Figure 5-6. *Getting to the sample experiment called Regression: Demand estimation*

2. Click the **Regression: Demand estimation** experiment. Figure 5-7 shows the experiment and the data used in the experiment.

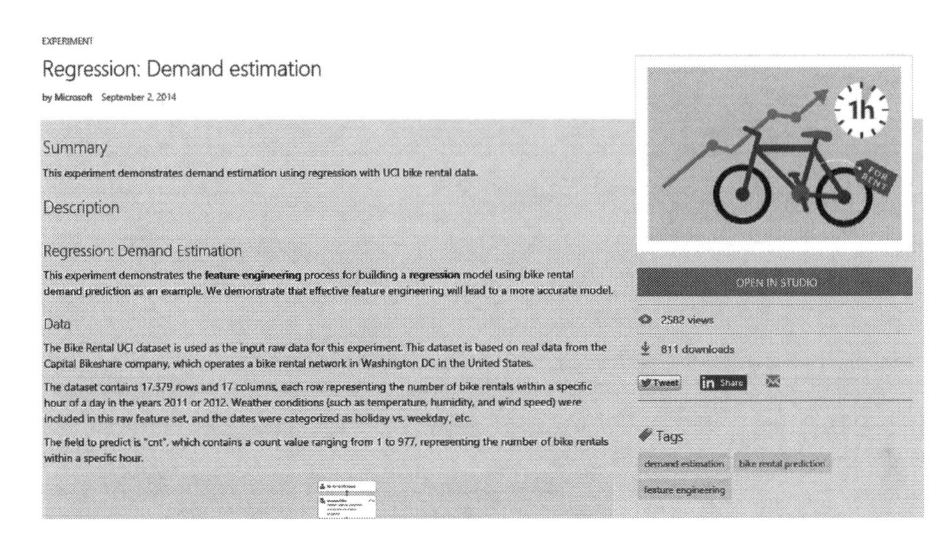

Figure 5-7. *The Regression: Demand estimation experiment*

3. Click **Open in Studio** to view the experiment in ML Studio (see Figure 5-8).

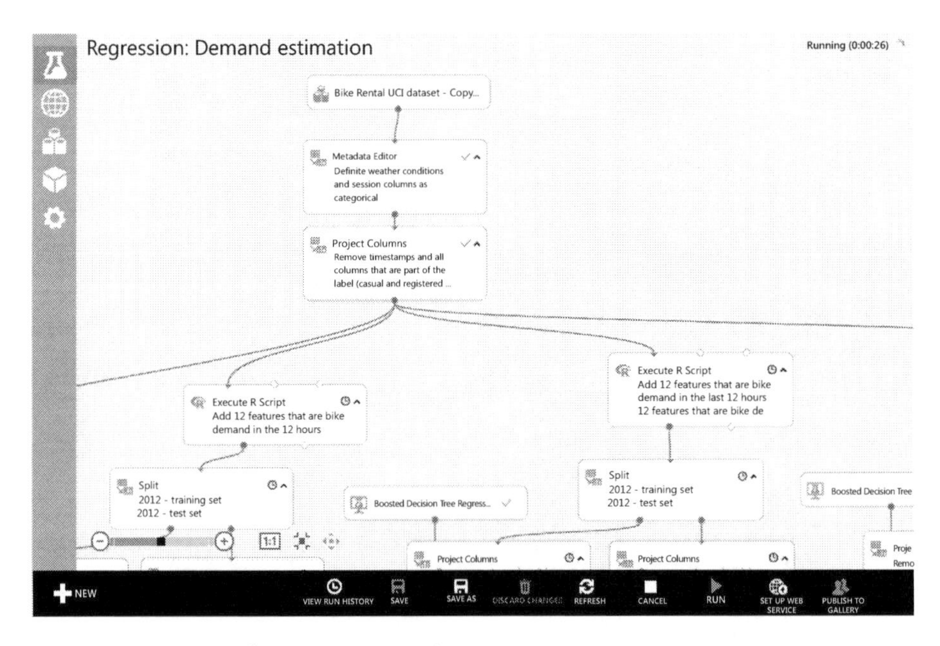

Figure 5-8. *Running the regression experiment*

■ **Note** If you have an Azure subscription, you can create a Machine Learning workspace using `https://manage.windowsazure.com`.

After you open the experiment in ML Studio, you can copy the experiment into the workspace you have created. To view the experiment, you can use the 8-hour guest access option by clicking **GUEST ACCESS** when entering ML Studio.

If you are using GUEST ACCESS, you will not be able to run the experiment, and you will have to sign in using a valid Microsoft account.

---

4. After you open the experiment in ML Studio, run the experiment by clicking the **Run** button at the bottom of the ML Studio.
5. Suppose in this experiment you want to make use of Python for pre-processing the input data before using it as inputs to the rest of the ML modules. To do this, **right-click the node** for the **Bike Rental UCI Dataset- Copy**.
6. Click **Generate Data Access Code**, as shown in Figure 5-9.

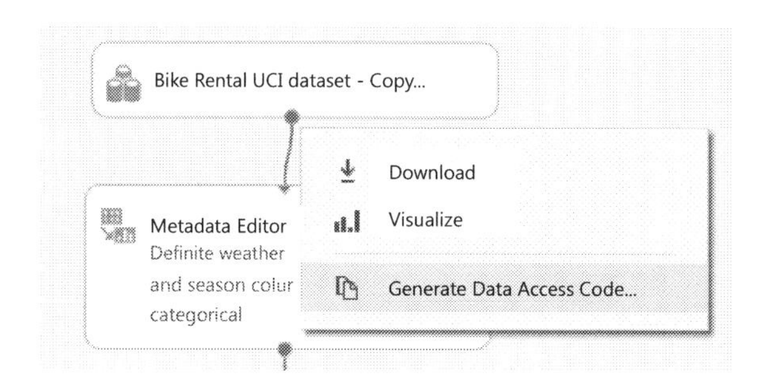

Figure 5-9. *Generating data access code*

■ **Note** The Generate Data Access Code option is accessible when you right-click a **dataset** or one of the **Data Format Conversion** modules (such as Convert to CSV). In all other Azure ML modules, this option is disabled.

---

7. This will generate the data access code in Python (shown in Figure 5-10). You can paste the data access code into a new IPython notebook.

The Python code contains the information needed to access the specific dataset and the **authorization_token**.

×

GENERATE DATA ACCESS CODE

Use this code to access your data

To programmatically access this dataset, simply copy the code snippet into your favorite development environment. Learn More

> Note: This code includes your workspace access token, which provides full access to your workspace. It should be treated like a password.

CODE SNIPPET

Python

```
from azurem1 import workspace
ws = Workspace(
     workspace_id='90840c6ddf1e4d468fc6db1d64f7af8b',
     authorization_token 'de3ec809e3394b3395538674c29d4069'
)
ds = ws.datasets[ ' Bike Rental UCI dataset - Copy (2) ' ]
frame = ds.to_dataframe()
```

☐ USE SECONDARY TOKEN

Figure 5-10. *Data Access Code*

8. Copy and paste the Python code into the IPython notebook, as shown in Figure 5-11.

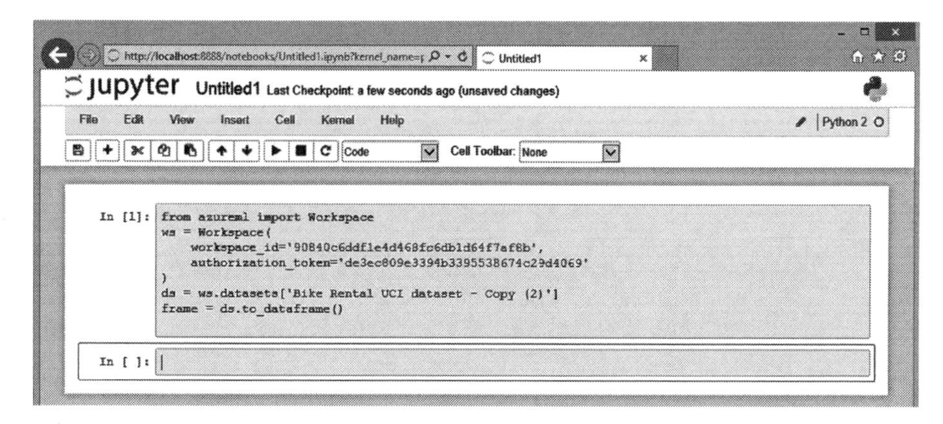

Figure 5-11. Using IPython notebook to access the dataset

9. In the Python code snippet that you have pasted, you can see that you are accessing the dataset in the workspace by name, where you specified the name of the dataset as **Bike Rental UCI dataset - Copy(2)**.

   After you have the data, you converted it into a pandas dataframe using the **ds.to_dataframe( )** function.

■ **Note** You can find out more about the internals of the AzureML Python library in the GitHub repository at `https://github.com/Azure/Azure-MachineLearning-ClientLibrary-Python`.

---

10. In the next IPython notebook cell, type **frame**, and click **Run Cell** (as shown in Figure 5-12).

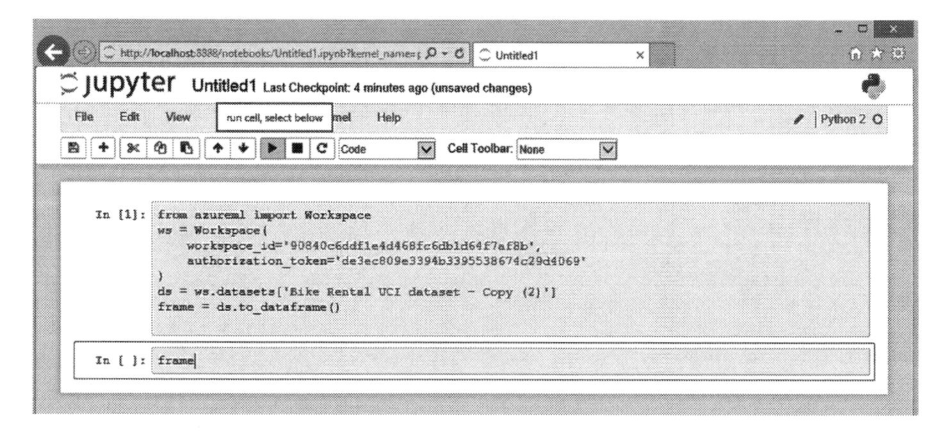

Figure 5-12. Running the data access code

11. After you run the selected cell, you will see the data contained within the pandas dataframe, as shown in Figure 5-13.

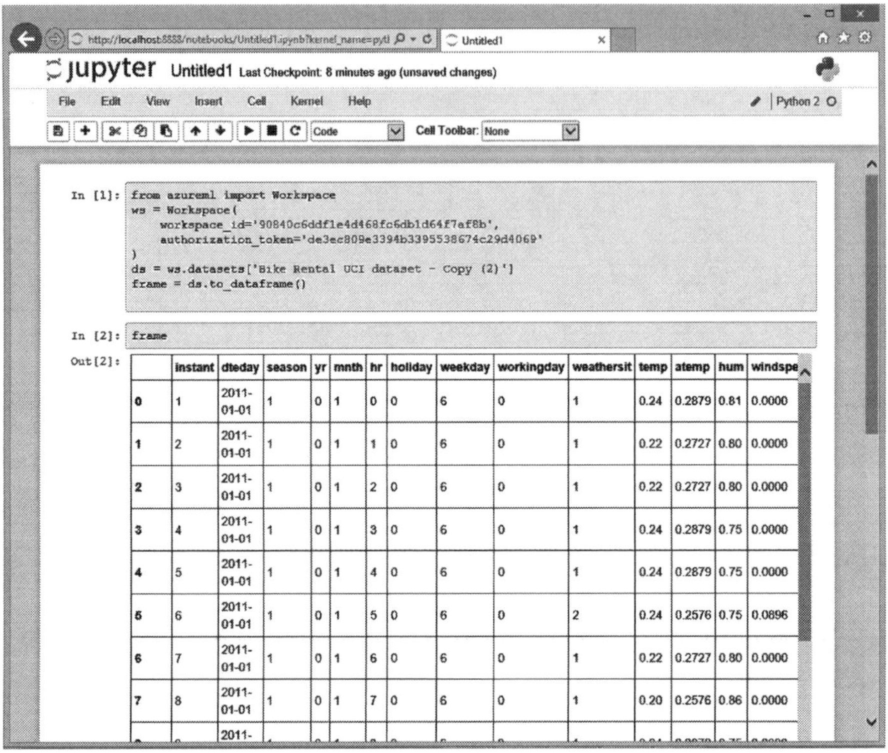

Figure 5-13. *Viewing the content of the dataframe*

Visualizations play an important role in data analysis and understanding the data that you are working with. Python provides powerful visualization capabilities that enable you to visualize data. Some of these visualization packages include matplotlib, VisPy, Bokeh, and ggplot for Python. In addition, pandas provides various plotting methods that can be used for visualizing the data in a pandas dataframe. These methods wrap around the basic plotting primitives provided by matplotlib.

Using the data you loaded in the earlier steps, let's plot the data to understand wind speed over time.

12. Cut and paste the following Python code in IPython notebook:

```
%pylab inline
frame.plot('dteday','windspeed', title='Windspeed
Over Time')
```

The first line specifies that the figures generated by the plot method are shown inline within the notebook.

The second line shows how to use the plot method provided by the pandas frame to plot the required figure, where the x-axis and y-axis are data from the `dteday` and `windspeed` columns, respectively.

Figure 5-14 shows the output from the plot.

Out[4]: <matplotlib.axes._subplots.AxesSubplot at 0xb2952e8>

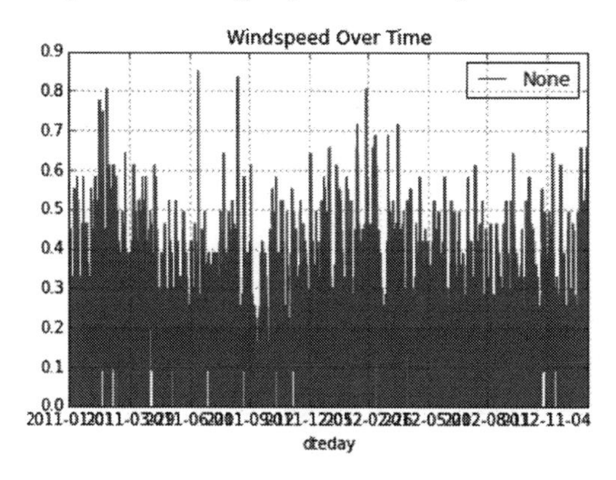

Figure 5-14. *Using the pandas dataframe plot method*

■ **Note** pandas provides various plotting functions. For a list of plotting capabilities supported, visit the pandas documentation page at `http://pandas.pydata.org/pandas-docs/stable/visualization.html`.

Inspired by a 2013 article on data visualization, the folks at the Data Science Lab show how to achieve beautiful visualization using matplotlib and pandas at `https://datasciencelab.wordpress.com/2013/12/21/beautiful-plots-with-pandas-and-matplotlib/`.

Using Python for Data Preprocessing

As you start building experiments using Azure ML, you might be bringing in raw data from various data sources. In order to use it in your experiments, you will need to clean and transform the raw data before it can be useful. For example, you might normalize the data, add additional columns, and combine data from several data sources. This process is often referred to as data wrangling.

Combining Data using Python

Python empowers you to perform data wrangling easily. More importantly, you can use IPython notebook to work with the data, experimenting with it before using the Python code in your experiment. In this section, you will learn how to do that. Specifically, you will learn how to use pandas for performing data wrangling. You will use two dataset samples provided in ML Studio: CRM Dataset Shared and CRM Appetency Labels.

1. After you have created a new experiment, search for the CRM Dataset Samples. You can find the CRM datasets by typing in the keyword **CRM** in the search box, as shown in Figure 5-15.

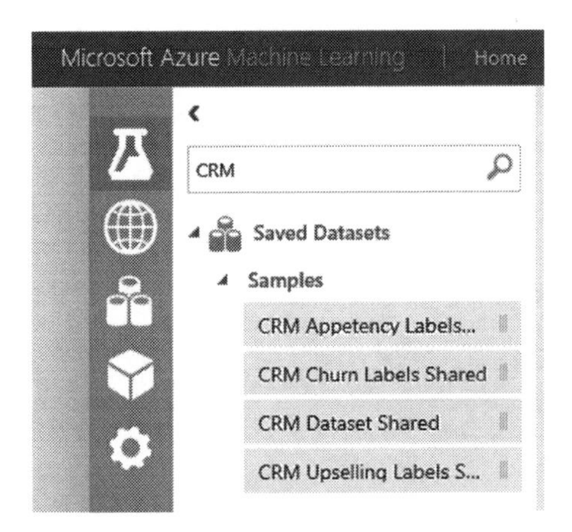

Figure 5-15. *CRM Dataset Samples*

2. Drag and drop the **CRM Dataset Shared** and **CRM Appetency Labels Shared** datasets into the experiment (shown in Figure 5-16).

Figure 5-16. *Using the CRM Dataset Samples in the experiment*

3. Similar to the steps that you performed in the earlier section, you can **right-click the node of any of the datasets** to **generate data access code**.

   Figure 5-17 shows how to right-click the node for **CRM Dataset Shared** to get the Python code needed to access the workspace.

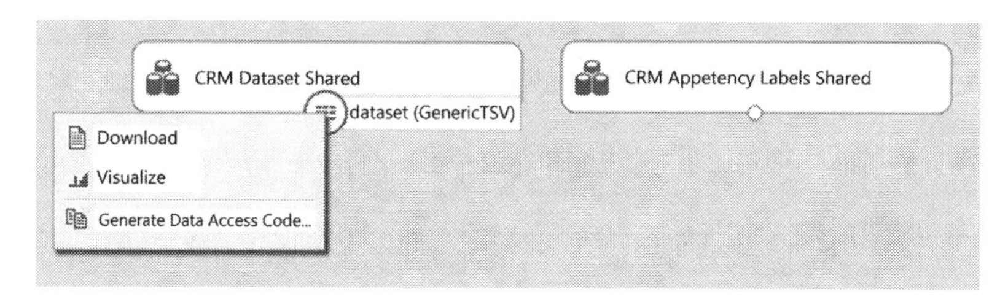

Figure 5-17. *Generating data access code*

4. Cut and paste the data access code (shown in Figure 5-18) into the IPython notebook.

GENERATE DATA ACCESS CODE

To programmatically access this dataset, simply copy the code snippet into your favorite development environment. Learn More.

Note: This code includes your workspace access token, which provides full access to your workspace. It should be treated like a password.

CODE SNIPPET

Python

```
from azureml import workspace
ws = Workspace(
    workspace_id='90840c6ddf1e4d468fc6db1d64f7af8b',
    authorization_token 'de3ec809e3394b3395538674c29d4069'
)
ds = ws.datasets[ ' CRM Dataset Shared ' ]
frame = ds.to_dataframe()
```

☐ USE SECONDARY TOKEN

Figure 5-18. *Data access code for the dataset called CRM Dataset Shared*

5. Next, specify the other Appetency dataset and load it into a dataframe.

   In IPython notebook, modify the Python code as follows:

```
from azureml import Workspace
ws = Workspace(
    workspace_id='<replace this with the workspace_id>',
    authorization_token=
         '<replace this with the authoritzation token>'
)

ds1 = ws.datasets['CRM Dataset Shared']
DatasetShareddf = ds1.to_dataframe()

ds2 = ws.datasets['CRM Appetency Labels Shared']
AppetencyLabeldf = ds2.to_dataframe()
```

   After you have modified the Python code, you can count the number of rows in each of the dataframes using the `len()` function.

   Figure 5-19 shows the number of rows in each of the dataframes.

```
In [15]: len(DatasetShareddf.index)

Out[15]: 50000

In [14]: len(AppetencyLabeldf.index)

Out[14]: 50000
```

Figure 5-19. *Finding the number of rows in each of the dataframes*

6. You can use the pandas function `concat` to concatenate the two datasets. To concatenate the two datasets, type the following code into IPython notebook:

```
import pandas as pd
SharedandAppetencydf = pd.concat(
        [DatasetShareddf,
         AppetencyLabeldf],
         axis=1)
```

Figure 5-20 shows visually how the two dataframes (**DatasetShareddf** and **AppetencyLabeldf**) are concatenated to produce the new dataframe called **SharedandAppetencydf**.

DatasetShareddf

	...	Var221	Var222	Var223	Var224	Var225	Var226	Var227	Var228	Var229	Var230
0	...	oslk	fXVEsaq	jySVZNlOJy	NaN	NaN	xb3V	RAYp	F2FyR07IdsN7I	NaN	NaN
1	...	oslk	2Kb5FSF	LM8l689qOp	NaN	NaN	fKCe	RAYp	F2FyR07IdsN7I	NaN	NaN
2	...	Al6ZaUT	NKv4yOc	jySVZNlOJy	NaN	kG3k	Qu4f	02N6s8f	ib5G6X1eUxUn6	am7c	NaN
3	...	oslk	CE7uk3u	LM8l689qOp	NaN	NaN	FSa2	RAYp	F2FyR07IdsN7I	NaN	NaN
4	...	oslk	1J2cvxe	LM8l689qOp	NaN	kG3k	FSa2	RAYp	F2FyR07IdsN7I	mj86	NaN

AppetencyLabeldf

	0
0	-1
1	-1
2	-1
3	-1
4	-1

SharedandAppetencydf

	...	Var222	Var223	Var224	Var225	Var226	Var227	Var228	Var229	Var230	0
0	...	fXVEsaq	jySVZNlOJy	NaN	NaN	xb3V	RAYp	F2FyR07IdsN7I	NaN	NaN	-1
1	...	2Kb5FSF	LM8l689qOp	NaN	NaN	fKCe	RAYp	F2FyR07IdsN7I	NaN	NaN	-1
2	...	NKv4yOc	jySVZNlOJy	NaN	kG3k	Qu4f	02N6s8f	ib5G6X1eUxUn6	am7c	NaN	-1
3	...	CE7uk3u	LM8l689qOp	NaN	NaN	FSa2	RAYp	F2FyR07IdsN7I	NaN	NaN	-1
4	...	1J2cvxe	LM8l689qOp	NaN	kG3k	FSa2	RAYp	F2FyR07IdsN7I	mj86	NaN	-1

Figure 5-20. *Concatenating dataframes*

■ **Note** Python provides powerful functions for merging, concatenating, and combining data in the pandas objects (such as DataFrame, Series, and Panel). These functions enable you to perform a join between two datasets (similar to how you perform a join in a relational database), using values from one dataset to patch the values in another dataset.

In this section, you learn how to use Python for concatenating dataframes (for illustration purpose only). You can also use the Add Columns module in ML Studio to concatenate two datasets.

Refer to `http://pandas.pydata.org/pandas-docs/stable/merging.html` for more examples.

Handling Missing Data Using Python

In this section, you'll continue using Python for other data wrangling tasks. Specifically, you will use Python to fill in missing values for specific columns, and perform data discretization on the data in some of the columns. You will be using the dataframe that you created in the previous section.

If you have not done that yet, type the following code into IPython notebook. The code uses the azureml module to access the datasets in the ML workspace, and convert them to a dataframe. In Listing 5-1, the `workspace_id` and `authorization_id` will be different for your experiments, and you will need to update it with the `workspace_id` and `authorization_id` that is provided in the data access code.

Listing 5-1. Complete Code for Accessing the ML Workspace

```
import pandas as pd

from azureml import Workspace
ws = Workspace(
    workspace_id='90840c6ddf1e4d468fc6db1d64f7af8b',
    authorization_token='de3ec809e3394b3395538674c29d4069'
)

ds1 = ws.datasets['CRM Dataset Shared']
DatasetShareddf = ds1.to_dataframe()

ds2 = ws.datasets['CRM Appetency Labels Shared']
AppetencyLabeldf = ds2.to_dataframe()
AppetencyLabeldf.rename(columns={0:'target'}, inplace=True)
SharedandAppetencydf = pd.concat(
                        [DatasetShareddf, AppetencyLabeldf],
                        axis=1)
```

Once the `SharedandAppetencydf` is created, start replacing the missing values. In Python, NaN is the default value used to denote missing data. In this section, you will replace NaN with the value 0. To do this, type the following code in IPython:

```
df3 = SharedandAppetencydf.fillna(0)
df3[0:5]
```

You use the `fillna()` method replace the NaN values with the value 0.

Figure 5-21 shows the first five rows of the datasets after performing `fillna()`. Notice that NaN has been replaced by 0.

```
In [56]: SharedandAppetencydf[0:5]
```

Out[56]:

	Var1	Var2	Var3	Var4	Var5	Var6	Var7	Var8	Var9	Var10	...
0	NaN	NaN	NaN	NaN	NaN	1526	7	NaN	NaN	NaN	...
1	NaN	NaN	NaN	NaN	NaN	525	0	NaN	NaN	NaN	...
2	NaN	NaN	NaN	NaN	NaN	5236	7	NaN	NaN	NaN	...
3	NaN	NaN	NaN	NaN	NaN	NaN	0	NaN	NaN	NaN	...
4	NaN	NaN	NaN	NaN	NaN	1029	7	NaN	NaN	NaN	...

5 rows × 231 columns

```
In [57]: df3 = SharedandAppetencydf.fillna(0)
         df3[0:5]
```

Out[57]:

	Var1	Var2	Var3	Var4	Var5	Var6	Var7	Var8	Var9	Var10	...
0	0	0	0	0	0	1526	7	0	0	0	...
1	0	0	0	0	0	525	0	0	0	0	...
2	0	0	0	0	0	5236	7	0	0	0	...
3	0	0	0	0	0	0	0	0	0	0	...
4	0	0	0	0	0	1029	7	0	0	0	...

5 rows × 231 columns

Figure 5-21. *Using fillna() on a dataframe*

■ **Note** For some datasets, exercise caution when replacing a value with 0. This might cause actual rows that contain 0 to become ineffective. Consider adding some constant value (such as 1) to the original dataset before replacing the NaN values.

Feature Selection Using Python

Feature selection (also known as variable selection) is frequently used by many data scientists to figure out the relevant features that will be useful inputs to an experiment.

By reducing the number of features used, the training time of an experiment can be significantly improved, and it often helps to reduce possibility of overfitting. Most importantly, it enables the model to be simplified, and leads to an easier understanding of the key features that contribute to a prediction.

The Python scikit-learn library provides powerful and comprehensive tools for pre-processing the data and performing model selection, feature selection, dimensionality reduction, and a rich set of machine learning algorithms (including classification, regression, and clustering). The NumPy library empowers Python developers with efficient N-dimensional array capabilities and a rich set of transformation capabilities (Fourier transforms, random number generation, and much more).

In this section, you will learn how to use the dataframes that you defined in the earlier sections of this chapter to determine the most relevant features of the datasets.

Many of the feature selection algorithms require numeric inputs, and it is important to figure out the data types specified in the dataframe. Let's start by understanding the data types for the data in the dataframe.

1. Assuming that you have already defined the dataframe SharedandAppetencydf, type the following code into IPython notebook:

```
dtypeGroups = df3.columns.to_series().groupby(df3.dtypes).groups
              list(dtypeGroups)
```

2. After you run the Python code, you will see the list of data types that are used in the dataframe (as shown in Figure 5-22). You can see the different dtypes used, including O (Object), int64, and float64.

```
In [108]: dtypeGroups = df3.columns.to_series().groupby(df3.dtypes).groups
          list(dtypeGroups)

Out[108]: [dtype('O'), dtype('int64'), dtype('float64')]
```

Figure 5-22. *Data types used in the dataframe*

■ **Note** pandas supports the use of the following dtypes: float, int, bool, datetime64[ns], timedelta[ns], and object. Each of the numeric dtypes has the item size specified (such as int64, float64). To understand more about pandas dtypes, refer to `http://pandas.pydata.org/pandas-docs/stable/basics.html#dtypes`.

---

3. As a preparation step for feature selection, do the following:

    a. Select the columns that contain numeric data.

    b. Specify the labels and features used for feature selection. To do this, copy and paste the following code into the IPython window:

```
numericdf = df3.select_dtypes(include=['float64'])
labels = df3["target"].values
features = numericdf.values
X, y = features, labels
features.shape
```

Figure 5-23 shows the output from running the code. The `features.shape` shows you that there are 50,000 rows and 191 columns that are of type float64. You omit columns of type int64 as features because the only int64 column is the `label` column.

```
In [111]: numericdf = df3.select_dtypes(include=['float64'])
          labels = df3["target"].values
          features = numericdf.values
          X, y = features, labels
          features.shape
Out[111]: (50000L, 191L)
```

Figure 5-23. *Defining the features and labels used for feature selection*

4. With the labels and features defined, you are ready to perform feature selection. Various feature selection techniques (such as univariate, L1-based, and tree-based feature selection) are provided in the scikit-learn library. You will use the tree-based feature selection to identify the important features. To do this, type the following code into the IPython notebook:

```
from sklearn.ensemble import ExtraTreesClassifier

model = ExtraTreesClassifier()
X_new = model.fit(X, y).transform(X)
model
```

Figure 5-24 shows the model output.

```
In [114]: from sklearn.ensemble import ExtraTreesClassifier

          model = ExtraTreesClassifier()
          X_new = model.fit(X, y).transform(X)
          model

Out[114]: ExtraTreesClassifier(bootstrap=False, compute_importances=None,
                     criterion='gini', max_depth=None, max_features='auto',
                     max_leaf_nodes=None, min_density=None, min_samples_leaf=1,
                     min_samples_split=2, n_estimators=10, n_jobs=1, oob_score=False,
                     random_state=None, verbose=0)
```

Figure 5-24. *Using tree-based feature selection*

5. Type the following into the IPython notebook to see the number of features selected:

```
X_new.shape
```

Figure 5-25 shows that there are 38 columns selected.

```
In [44]: X_new.shape

Out [44]: (50000L, 38L)
```

Figure 5-25. *Number of features selected*

6. Next, you will use the information on the selected features, and extract the relevant columns from the dataframe df3 (which you defined in an earlier step).

Type the following code into IPython notebook:

```
import NumPy as np
importances = model.feature_importances_

idx = np.arange(0, X.shape[1])
features_to_keep =
        idx[importances > np.mean(importances)]

df4 = df3.ix[:,features_to_keep]
df5 = pd.concat([df4, AppetencyLabeldf], axis=1)
df5[0:5]
```

After the code is run, you will see the updated dataframe with 39 columns. The last column is the `label` column.

Figure 5-26 shows the output from running the Python code.

```
In [117]: df5[0:5]
Out[117]:
```

	Var6	Var7	Var13	Var21	Var22	Var24	Var25	Var28	Var35	Var38	...
0	1526	7	184	464	580	14	128	166.56	0	3570	...
1	525	0	0	168	210	2	24	353.52	0	4764966	...
2	5236	7	904	1212	1515	26	816	220.08	0	5883894	...
3	0	0	0	0	0	0	0	22.03	0	0	...
4	1029	7	3216	64	80	4	64	200.00	0	0	...

5 rows × 39 columns

Figure 5-26. *Defining the dataframe, after feature selection*

Congratulations! You have successfully performed feature selection on the dataset that you are using in ML studio. IPython notebook provides you with an interactive environment where you can explore data, perform data wrangling, and get the dataset into a shape ready to be used in the subsequent steps in an experiment in ML studio. In the next section, you will learn how to use this in ML Studio.

Running Python Code in an Azure ML Experiment

In the earlier sections, you learned how to access the datasets in your Machine Learning workspace using IPython notebook. The complete code is shown in Listing 5-2.

Listing 5-2. Complete Python Code for the Experiment

```
import NumPy as np
import pandas as pd
from sklearn.ensemble import ExtraTreesClassifier
from azureml import Workspace
ws = Workspace(
    workspace_id='<workspace id>',
    authorization_token='<authorization token>'
)

ds1 = ws.datasets['CRM Dataset Shared']
DatasetShareddf = ds1.to_dataframe()

ds2 = ws.datasets['CRM Appetency Labels Shared']
AppetencyLabeldf = ds2.to_dataframe()
AppetencyLabeldf.rename(columns={0:'target'}, inplace=True)

SharedandAppetencydf = pd.concat(
                         [DatasetShareddf, AppetencyLabeldf],
                         axis=1)
```

```
df3 = SharedandAppetencydf.fillna(0)

numericdf = df3.select_dtypes(include=['float64'])
labels = df3["target"].values
features = numericdf.values
X, y = features, labels

model = ExtraTreesClassifier()
X_new = model.fit(X, y).transform(X)

importances = model.feature_importances_
idx = np.arange(0, X.shape[1])
features_to_keep = idx[importances > np.mean(importances)]

x_feature_selected = X[:, features_to_keep]
df4 = df3.ix[:,features_to_keep]
df5 = pd.concat([df4, AppetencyLabeldf], axis=1)
```

To use the Python code that you have developed and tested in IPython notebook, use the Execute Python Script module to execute the Python script. During execution, the **Execute Python Script module** runs on a backend that uses Anaconda 2.1. Because the Python runtime runs in a sandbox, the Python script will not be able to access the network and the local file system on the machine where it is running.

You can either provide the Python script within the module, or you can use a script bundle (similar to how you used a script ZIP bundle when executing an R script). If you have additional Python modules that you plan to use, you should include the additional Python files as part of the ZIP bundle.

When using the Execute Python Script module, the module requires that you specify an entry-point function called **azureml_main**, and return a single data frame. In order to run the code in Listing 5-3 using the Execute Python Script module, you need to make the following changes:

- You must change the names of the input dataframes. The names of the dataframe for the Execute Python Code module are dataframe1 and dataframe2.
- Depending on the version of Anaconda that you have installed on your local development machine, some functions might not work when running in an Azure ML experiment.

- For example, you can use the select_dtypes function locally, but when used in the Execute Python Script module, an exception is thrown. So change the line

  ```
  numericdf =
    df3.select_dtypes(include=['float64'])
  ```

  to

  ```
  numericdf =
        df3.loc[:, df3.dtypes == np.float64]
  ```

Listing 5-3. Modified Python Code Used in the Azure ML Experiment

```
# The script MUST contain a function named azureml_main
# which is the entry point for this module.
#
# The entry point function can contain up to two input arguments:
#   Param<dataframe1>: a pandas.DataFrame
#   Param<dataframe2>: a pandas.DataFrame
def azureml_main(dataframe1, dataframe2):

    # Execution logic goes here
    import NumPy as np
    import pandas as pd
    from sklearn.ensemble import ExtraTreesClassifier

    dataframe2.rename(columns={'Col1':'target'}, inplace=True)
    SharedandAppetencydf = pd.concat(
                        [dataframe1, dataframe2],
                        axis=1)

    df3 = SharedandAppetencydf.fillna(0)
    numericdf = df3.loc[:, df3.dtypes == np.float64]

    labels = df3["target"].values
    features = numericdf.values
    X, y = features, labels

    model = ExtraTreesClassifier()
    X_new = model.fit(X, y).transform(X)

    print(X_new.shape)

    importances = model.feature_importances_

    idx = np.arange(0, X.shape[1])
    features_to_keep = idx[importances > np.mean(importances)]
```

```
    df4 = numericdf.ix[:,features_to_keep]
    print(df4.shape)

    df5 = pd.concat([df4, dataframe2], axis=1)
    return df5,
```

Let's use the modified code given in Listing 5-3 in an Azure ML experiment.

1. Using the experiment that you used earlier, drag and drop an **Execute Python Script** module.
2. Connect the **CRM Dataset Shared** dataset as the **first input dataset** to the **Execute Python Script** module.
3. Connect the CRM **Appetency Labels Shared** dataset as the **second input dataset** to the **Execute Python Script** module.
4. After you have done that, **copy and paste** the code in Figure 5-27 to the **Execute Python Script window**. Figure 5-28 shows the completed experiment.

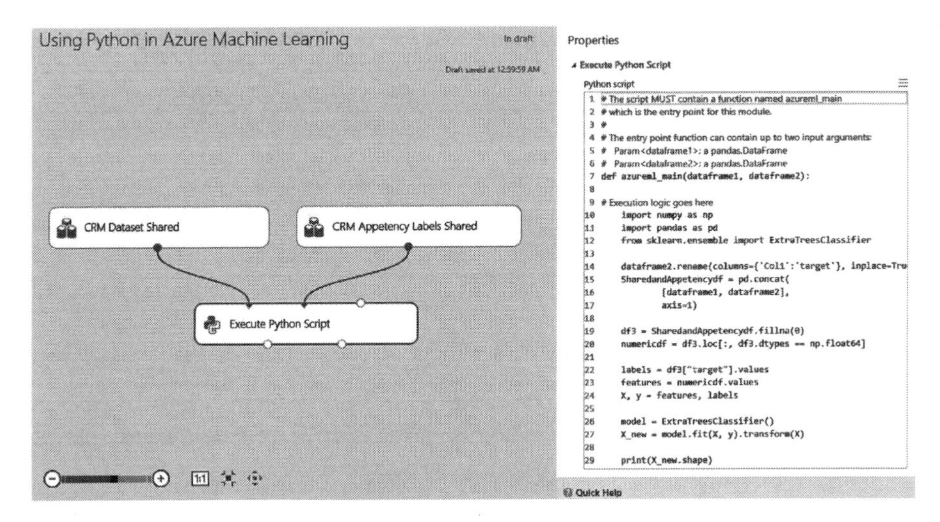

Figure 5-27. *Using the Execute Python Script*

Using Python in Azure Machine Learning › Execute Python Script › Results dataset

rows 50000 columns 5

Var28	Var57	Var81	Var113	Target
166.56	4.076907	7333.11	117625.6	-1
353.52	5.408032	151098.9	-356411.6	-1
220.08	6.599658	16211.58	405104	-1
22.08	1.98825	0	-275703.6	-1
200	4.552446	37423.5	10714.84	-1
200	0.166417	11370.72	369814	-1
176.56	5.448622	31655.4	-808528	-1
230.56	5.067507	402441	101923.6	-1
300.32	2.045717	117761.7	161314.4	-1
166.56	6.326853	149686.5	842480	-1
133.12	4.091647	2739.141	-4831.72	-1
133.12	4.432173	9885.09	67166.4	-1
240.56	1.899594	34917	-45625.2	-1
176.56	2.524888	19991.43	-179277.6	-1

Figure 5-28. *Results dataset after performing feature selection using Python*

You are now ready to run the Azure ML experiment, which uses a Python script for feature selection.

5. Click **Run** to start the experiment
6. After the experiment has successfully run, you can right-click the **Results dataset node** to see the features that have been selected. Figure 5-28 shows the four features selected.
7. You can also right-click the **Python device node** to see the output generated when the experiment has completed running the Execute Python Script module. Figure 5-29 shows the output.

Using Python in Azure Machine Learning › Execute Python Script › Python device

◢ Standard Output

```
Microsoft Drawbridge Console Host [Version 1.0.2108.0]
[Error 2] The system cannot find the file specified (ignoring)
Arg Index [0] = [C:\server\invokepy.py]
Arg Index [1] = [--batch]
Arg Index [2] = [C:\temp]
Arg Index [3] = [azuremod]
Arg Index [4] = [2]
Arg Index [5] = [461b4992bbba4af695659c3aa6051e02.xdr]
Arg Index [6] = [2015e1521a884eac88293b25e24d6c35.xdr]
Arg Index [7] = [1]
Arg Index [8] = [.maml.oport1]
Started in [C:\temp]
Running in [C:\temp]
Executing azuremod with inputs ['461b4992bbba4af695659c3aa6051e02.xdr',
'2015e1521a884eac88293b25e24d6c35.xdr'] and generating outputs ['.maml.oport1']
(50000L, 4L)
(50000, 4)
```

◢ Standard Error

```
Interpreter produced no output.
```

◢ Graphics

Figure 5-29. *Python device output*

Summary

In this chapter, you learned about the exciting possibilities and scenarios that are enabled by the integration of Python into Azure Machine Learning execute Python scripts from within an ML experiment. Several commonly used Python libraries (pandas, scikit-learn, matplotlib, and NumPy) are available as part of the latest distribution.

You learned how to use these libraries to transform the data from raw data to a finished form. In addition, you learned how to perform data processing, and fill in missing values using IPython notebook. You learned how to integrate Python code that you have developed and tested using IPython notebook, and integrate it as part of an Azure Machine Learning experiment using the Execute Python Script module.

PART II

Statistical and Machine Learning Algorithms

CHAPTER 6

Introduction to Statistical and Machine Learning Algorithms

This chapter will serve as a reference for some of the most commonly used algorithms in Microsoft Azure Machine Learning. We will provide a brief introduction to algorithms such as linear regression, k-means for clustering, decision trees, boosted decision trees, neural networks, support vector machines, and Bayes point machines.

This chapter will provide a good foundation and reference material for some of the algorithms you will encounter in the rest of the book. We group these algorithms into the following categories:

- Regression
- Classification
- Clustering

Regression Algorithms

Let's first talk about the commonly used regression techniques in the Azure Machine Learning service. Regression techniques are used to predict response variables with numerical outcomes, such as predicting the miles per gallon of a car or predicting the temperature of a city. The input variables may be numeric or categorical. However, what is common with these algorithms is that the output (or response variable) is numeric. We'll review some of the most commonly used regression techniques including linear regression, neural networks, decision trees, and boosted decision tree regression.

Linear Regression

Linear regression is one of the oldest prediction techniques in statistics. In fact, it traces its roots back to the work of Carl Friedrich Gauss in 1795. The goal of linear regression is to fit a linear model between the response and independent variables, and use it to predict the outcome given a set of observed independent variables. A simple linear regression model is a formula with the structure of

$$Y = \beta_0 + \beta_1 X_1 + \beta_2 X_2 + \beta_3 X_3 + \beta_4 X_4 + \cdots + \varepsilon$$

where

- Y is the response variable (the outcome you are trying to predict), such as miles per gallon.
- X_1, X_2, X_3, etc. are the independent variables used to predict the outcome.
- β_0 is a constant that is the intercept of the regression line.
- β_1, β_2, β_3, etc. are the coefficients of the independent variables. These refer to the partial slopes of each variable.
- ε is the error or noise associated with the response variable that cannot be explained by the independent variables X1, X2, and X3.

A linear regression model has two components: a deterministic portion (i.e. $\beta_1 X_1 + \beta_2 X_2 + \cdots$) and a random portion (i.e. the error, ε). You can think of these two components as the signal and noise in the model.

If you only have one input variable, X, the regression model is the best line that fits the data. Figure 6-1 shows an example of a simple linear regression model that predicts a car's miles per gallon from its horsepower. With two input variables, the linear regression is the best plane that fits a set of data points in a 3D space. The coefficients of the variables (β_1,β_2,β_3, etc.) are the partial slopes of each variable. If you hold all other variables constant, then the outcome Y will increase by β_1 when the variable X_1 increases by 1. This is why economists typically use the phrase "*ceteris paribus*" or "*all other things being equal*" to describe the effect of one independent variable on a given outcome.

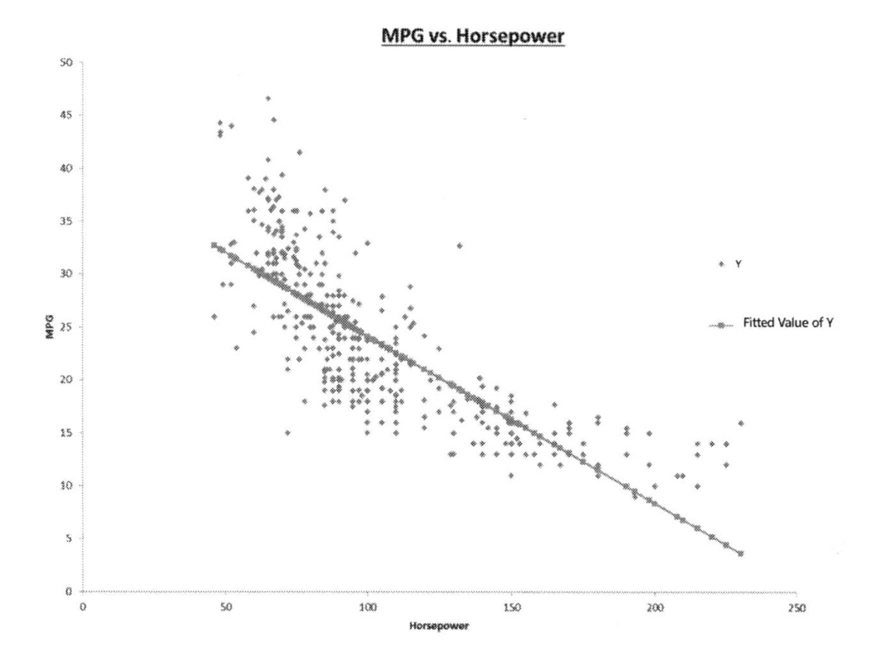

Figure 6-1. *A simple linear regression model that predicts a car's miles per gallon from its horsepower*

Linear regression uses the least squares or gradient descent methods to find the best model coefficients for a given dataset. The least squares method achieves this by minimizing the sum of the squared error between the fitted and actual values of each observation in the training data. The gradient descent method finds the optimal model coefficients by updating the coefficients in each iteration. Each update ensures that the sum of errors between the model fitted and the actual values of the training data are reduced. Through several iterations it will find the local minimum by moving in the direction of the negative gradient, hence the name.

▒ **Note** You can learn more about linear regression from the book *Business Analysis Using Regression: A Casebook* by David P. Foster, Robert A. Stine, and Richard P. Waterman (Springer-Verlag, 1998).

Neural Networks

Artificial neural networks are a set of algorithms that mimic the functioning of the brain. There are many different neural network algorithms, including backpropagation networks, Hopfield networks, Kohonen networks (also known as self-organizing maps), and adaptive resonance theory (or ART) networks. The most common is the back-propagation algorithm, also known as multilayered perception.

The back-propagation network has several neurons arranged in layers. The most commonly used architecture is the three-layered network shown in Figure 6-2. This architecture has one input, one hidden layer, and one output layer. However, you can also have two or more hidden layers. The number of input and output nodes are determined by the dataset. Basically, the number of input nodes equals the number of independent variables you want to use to predict the output. The number of output nodes is the same as the number of response variables. In contrast, the number of hidden nodes is more flexible.

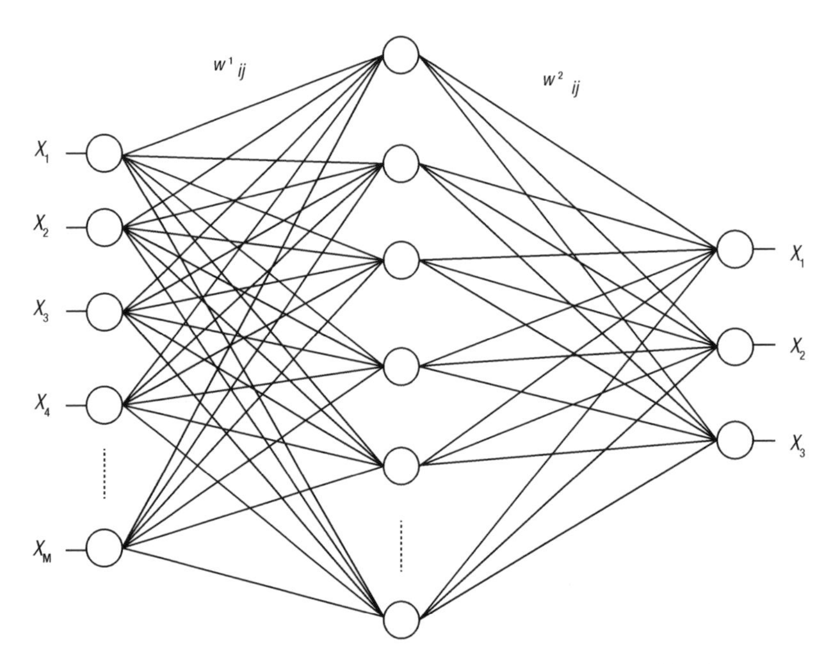

Figure 6-2. *A neural network with three layers: one input, one hidden layer, and one output layer*

The development of a neural network model is carried out in two steps: training and testing. During training, you show the neural network a set of examples from the training set. Each example has values of the independent as well as the response variables. During training, you show the examples several times to the neural network. At each iteration, the network predicts the response. In the forward propagation phase of training, each node in the hidden and output layers calculates a weighted sum of its inputs, and then uses this sum to compute its output through an activation function. The output of each neuron in the neural network usually uses the following sigmoidal activate function:

$$f(x) = \frac{1}{1+e^{-x}}$$

There are, however, other activation functions that can be used in neural networks, such as Gaussian, hyperbolic tangent (tanh), linear threshold, and even a simple linear function.

Let's assume there are M input nodes. The connection weights between the input nodes and the first hidden layer are denoted by w^1_{ij}.

At each hidden node the weighted sum is given by

$$S_j = \sum_{i=0}^{M-1} \left(a_i w^1_{ij} \right)$$

When the weighted sum is calculated, the sigmoidal activate function is calculated as follows:

$$f(S_j) = \frac{1}{1+e^{-S_j}}$$

After the activation level of the output node is calculated, the backward propagation step starts. In this phase, the algorithm calculates the error of its prediction based on the actual response value. Using the gradient descent method, it adjusts the weights of all connections proportion to the error. The weights are adjusted in a manner that reduces the error the next time this example is presented to the network. After several iterations, the neural network converges to a solution.

During testing, you simply use the trained model to score records. For each record, the neural network predicts the value of the response for a given set of input variables (see Figure 6-2).

The learning rate determines the rate of convergence to a solution. If the learning rate is too low, the algorithm will need more learning iterations (and hence more time) to converge to the minimum. In contrast, if the learning rate is too large, the algorithm could potentially bounce around and may never find the local minimum. Hence, the neural net will be a poor predictor.

Another important parameter is the number of hidden nodes. The accuracy of the neural network may increase with the number of hidden nodes. However, additional hidden nodes can increase the processing time and can lead to over-fitting. In general, increasing the number of hidden nodes or hidden layers can easily lead to over-parameterization, which will increase the risk of over-fitting. One rule of thumb is to start with the number of hidden nodes equal roughly to the square root of the number of input nodes. Another general rule of thumb is that the number of neurons in the hidden layer should be between the size of the input layer and the size of the output layer. For example, (number of input nodes + number of output nodes) x 2/3. These rules of thumb are merely starting points, intended to avoid over-fitting; the optimal number can only be found through experimentation and validation of performance on test data.

Decision Trees

Decision tree algorithms are hierarchical techniques that work by splitting the dataset iteratively based on certain statistical criteria. The goal of decision trees is to maximize the variance across different nodes in the tree and minimize the variance within each node. Figure 6-3 shows a simple decision tree created with two splits of the data. The root node (Node 0) contains all the data in the dataset. The algorithm splits the data based on a defined statistic, creating three new nodes (Node 1, Node 2, and Node 3). Using the same statistic, it splits the data again at Node 1, creating two more leaf nodes (Nodes 4 and 5). The decision tree makes its prediction for each data row by traversing to the leaf nodes (one of the terminal nodes: Node 2, 3, 4, or 5).

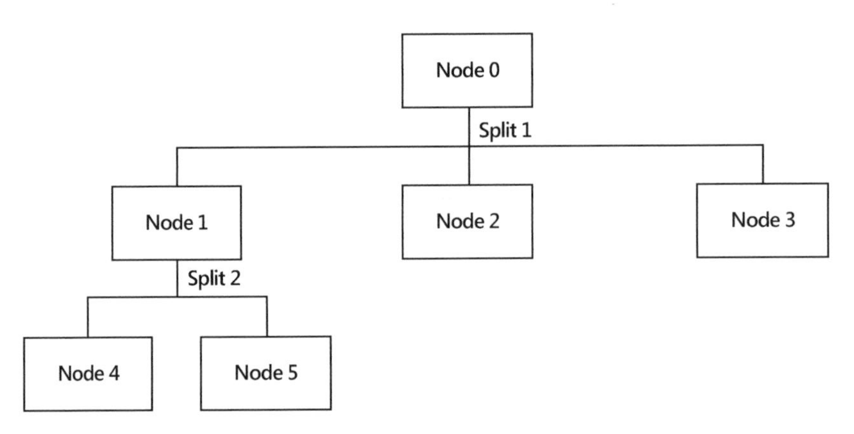

Figure 6-3. *A simple decision tree with two data splits*

Figure 6-4 shows a fictional example of a very simple decision tree that predicts whether a customer will buy a bike or not. In this example, the original dataset has 100 examples. The most predictive variable is age, so the decision first splits the data by age. Customers younger than 30 fall in the left branch while those aged 30 or above fall in the right branch. The next most important variable is gender, so in the next level the decision tree splits the data by gender. In the younger branch (for customer customers under 30), the decision tree splits the data into male and female branches. It also does the same for the older branch. Finally, Figure 6-4 shows the number of examples in each node and the probability to purchase. As a result, if you have a female customer aged 23, the tree predicts that she only has a 30% chance of buying the bike because she will end up in Node 3. A male customer aged 45 will have an 80% chance of buying a bike since he will end up in Node 6 in the tree.

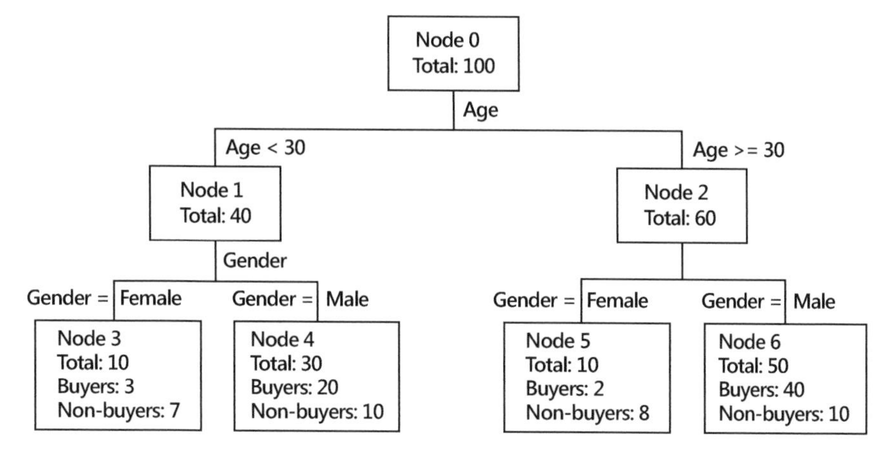

Figure 6-4. *A simple decision tree to predict likelihood to buy bikes*

Some of the most commonly used decision tree algorithms include Iterative Dichotomizer 3 (ID3), C4.5 and C5.0 (successors of ID3), Automatic Interaction Detection (AID), Chi-squared Automatic Interaction Detection (CHAID), and Classification and Regression Tree (CART). While very useful, the ID3, C4.5, C5.0, and CHAID algorithms are classification algorithms and are not useful for regression. As the name suggests, the CART algorithm can be used for either classification or regression.

How do you choose the variable to use for splitting data at each level? Each decision tree algorithm uses a different statistic to choose the best variable for splitting. ID3, C4.5, and C5.0 use information gain, while CART uses a metric called Gini impunity. Gini impunity measures the misclassification rate of a randomly chosen example.

■ **Note** More information on decision trees is available in the book *Data Mining and Market Intelligence for Optimal Market Returns* by S. Chiu and D. Tavella (Oxford, UK, Butterworth-Heinemann, 2008) and at `http://en.wikipedia.org/wiki/Decision_tree_learning`.

Boosted Decision Trees

Boosted decision trees are a form of ensemble models. Like other ensemble models, boosted decision trees use several decision trees to produce superior predictors. Each of the individual decision trees can be weak predictors. However, when combined they produce superior results.

As discussed in Chapter 1, there are three key steps to building an ensemble model: a) data selection, b) training classifiers, and c) combining classifiers.

The first step to build an ensemble model is data selection for the classifier models. When sampling the data, a key goal is to maximize diversity of the models, since this improves the accuracy of the solution. In general, the more diverse your models, the better the performance of your final classifier, and the smaller the variance of its predictions.

Step 2 of the process entails training several individual classifiers. But how do you assign the classifiers? Of the many available strategies, the two most popular are bagging and boosting. The bagging algorithm uses different random subsets of the data to train each model. The models can then be trained in parallel over their random subset of training data. In contrast, the boosting algorithm improves overall performance by sequentially training the models, testing the performance of each on the training data. The examples in the training set that were misclassified are given more importance in subsequent training. Thus, during training, each additional model focuses more on the misclassified data. The boosted decision tree algorithm uses the boosting strategy. In this case, every new decision tree is trained with emphasis on the misclassified cases to reduce the error rates. This is how a boosted decision tree produces superior results from weak decision trees.

It is important to watch two key parameters of the boosted decision tree: the number of leaves in each decision tree and the number of boosting iterations. The number of leaves in each tree determines the amount of interaction allowed between variables in the model. If this number is set to 2, then no interaction is allowed. If it is set to 3, the model can include effects of interaction of at most two variables. You need to try different values to find the one that works best for your dataset. It has been reported that 6-8 leaves per tree yields good results for most applications. In contrast, having only two leaves per tree leads to poor results. The second important parameter to tweak is the number of boosting iterations (the number of trees in the model). A very large number of trees reduces the errors, but easily leads to over-fitting. To avoid over-fitting, you need to find an optimal number of trees that minimizes the error on a validation dataset.

Finally, once you train all the classifiers, the last step is to combine their results to make a final prediction. There are several approaches to combining the outcomes, ranging from a simple majority to a weighted majority voting.

■ **Note** You can learn more about boosted decision trees from the book *Ensemble Machine Learning, Methods and Applications* by C. Zhang and Y. Ma (New York, NY: pp. 1 - 34, Springer, 2012) and at `http://en.wikipedia.org/wiki/Gradient_boosting#Gradient_tree_boosting`.

Classification Algorithms

Classification is a type of supervised machine learning. In supervised learning, the goal is to infer a function using labeled training data. The function can then be used to determine the label for a new dataset (where the labels are unknown). A non-exhaustive list of classification algorithms that can be used for building the model includes decision trees, logistic regression, neural networks, support vector machines, naïve Bayes, and Bayes Point Machines.

Classification algorithms are used to predict the label for input data (where the label is unknown). Labels are also referred to as classes, groups, or target variables. For example, a telecommunication company wants to predict the following:

- **Churn**: Customers who have an inclination to switch to a different telecommunication provider
- **Propensity to Buy**: Customers willing to buy new products or services
- **Upselling**: Customers willing to buy upgraded services or add-ons

To achieve this, the telecommunication company builds a classification model using training data (where the labels are known or have already been predefined). In this section, you'll look at several common classification algorithms that can be used for building the model. Once the model has been built and validated using test data, data scientists at the telecommunication company can use the model to predict churn, propensity to buy, and upselling labels for customers (where the labels are unknown). Consequently, the telecommunication company can use these predictions to design marketing strategies that can reduce the customer churn and offer services to the customers that are more willing to buy new services or upsell.

Other scenarios where classification algorithms are commonly used include financial institutions, where models are used to determine whether a credit card transaction is a fraudulent case or if a loan application should be approved based on the financial profile of the customer. Hotels and airlines use models to determine whether a customer should be upgraded to a higher level of service (such as from economy to business class, from a normal room to a suite, etc.).

The classification problem is defined as follows: given an input sample of $X = (x_1, x_2, \cdots, x_d)$, where x_1 refers to an item in the sample of size d, the goal of classification is to learn the mapping $X \to Y$, where $y \in Y$ is a class.

An instance of data belongs to one of J groups (or classes), such as $C_1, C_2, \cdots, C_j$. For example, in a two-class classification problem for the telecommunication scenario, Class C_1 refers to customers that will churn and switch to a new telecommunication provider, and Class C_2 refers to customers that will not churn.

To achieve this, labeled training data is first used to train a model using one of the classification algorithms. This is then validated by using test data to determine the number of mistakes made by the trained model (the classifier). Various metrics are used to measure the performance of the classifier. These include measuring the accuracy, precision, recall, and the area under curve (AUC) of the trained model.

In the earlier sections, you learned how decision trees, boosted decision trees, and neural networks work. These algorithms are useful for both regression and classification. In this section, you will learn how support vector machines and Bayes Point Machines work.

Support Vector Machines

Support vector machines (SVMs) were introduced by Bernhard E. Boser, Isabelle Guyon, and Vladimir N. Vapnik at the Conference on Learning Theory (COLOT) in 1992. A SVM is based on techniques grounded in statistical learning theory, and is considered a type of kernel-based learning algorithm.

The core idea of SVMs is to find the separating hyperplane that separates the training data into two classes, with a gap (or margin) that is as wide as possible. When there are only two input variables, a straight line separates the data into two classes. In higher-dimensional space (more than two input variables), a hyperplane separates the training data into two classes. Figure 6-5 shows how a hyperplane separates the two classes and the margin. The circled items are the support vectors because they define the optimal hyperplane, which provides the maximum margin.

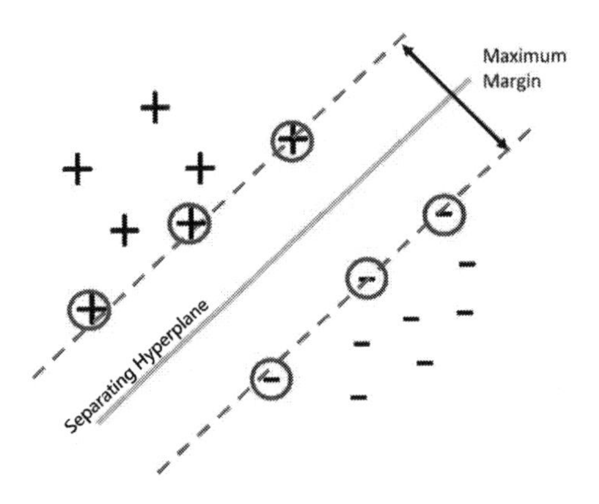

Figure 6-5. *Support vector machine, separating hyperplane, and margin*

Consider the following example. Suppose a telecommunication company has the following training data consisting of *n* customers. Out of these n customers, let's assume that 50 customers will churn, and the other 50 customers will not. For each customer, you extract 10 input variables (or features) that will be used to represent the customer. Given a customer who has used the service for some time, the data scientist and business analysts in the telecommunication company want to determine whether this customer will churn and move to a different telecommunication provider.

Suppose the training data consists of the following: $(x_1, y_1), \ldots, (x_n, y_n)$, where (x_i, y_j) denotes to x_i mapped to the class y_j. The hyperplane decision function is

$$D(x) = (w \cdot x) + w_0$$

where w and w_0 are coefficients. A separating hyperplane will satisfy the following constraints:

$$(w \cdot x) + w_0 \geq +1$$

if $y_i = +1$

$$(w \cdot x) + w_0 \leq -1 \quad \text{if } y_i = -1,\ i = 1,\ldots,n$$

An optimal separating hyperplane is one that enables the maximum margin between the two classes. These two constraints for describing the hyperplane can be represented using the equation

$$y_i\left[(w \cdot x)+w_0\right] \geq 1 \text{ where } i=1, \ldots, n .$$

This equation can represent all hyperplanes that can be used for separating the data. Often, the equation is not solved directly in its current form (also referred to as the primal form), due to the difficulty in directly computing the value of the norm of $\|w\|$. In practice, the dual form of the equation is used for solving the optimization problem that will identify the optimal hyperplane.

■ **Note** You can learn more about support vector machines and how the margin can be maximized from the book *Learning From Data (Concepts, Theory and Methods)* by Vladimir Cherkassky and Filip M. Mulier (Wiley-Interscience, 1998). A good overview of support vector machines can be found at `http://en.wikipedia.org/wiki/Support_vector_machine`.

Azure Machine Learning provides a Two-Class Support Vector Machine module, which enables you to build a model that supports binary predictions. The inputs to the module can be either continuous and/or categorical variables. The module provides several parameters that can be used to fine-tune the behavior of the support vector machine algorithm. These include

- **Number of iterations**: Determines the speed of training the model. This parameter enables you to balance between training speed and model accuracy. The default value is set to 1.
- **Lambda**: Weight for L1 regularization used for tuning the complexity of the model that is produced. The default value of Lambda is set to 0.001. A non-zero value is used to avoid over-fitting the model.
- **Normalize features**: Determines whether the algorithm normalizes the values.
- **Project to unit-sphere**: Determines whether the algorithm projects the values to a unit-sphere.
- **Random number seed**: The seed value used for random number generation when computing the model.
- **Allow unknown categorical levels**: Determines whether the algorithm supports unknown categorical values. If this is set to True, the algorithm creates an additional level for each categorical column. The additional level is used for mapping levels in the test dataset that are not found in the training dataset.

Bayes Point Machines

Bayes Point Machines (BPMs) are a type of linear classification algorithm that was introduced by Ralf Herbrich, Thore Grapel, and Colin Campbell in 2001. The core idea of the Bayes Point Machine algorithm is to identify an "average" classifier that is able to effectively and efficiently approximate the theoretical optimal Bayesian average of several linear classifiers (based on their ability to generalize). The "average" classifier is known as the Bayes point. In empirical studies, Bayes Point Machines have consistently outperformed support vector machines for both lab and real-world data.

■ **Note** The Bayes Point Machine algorithm used in Azure Machine Learning is based on Infert.Net and provides improvements to the original Bayes Point Machine algorithm. These improvements enable the Bayes Point Machine used in Azure Machine Learning to be more robust and less prone to over-fitting of the data. It also reduces the need to perform performance tunings.

Some of these improvements include the use of expectation propagation as the message-passing algorithm. In addition, the implementation does not require the use of parameter sweeping and having normalized data.

Recall the earlier definition of the classification problem: given an input sample of $X = (x_1, x_2, \cdots, x_d)$, where x_i refers to an item in the sample of size d, the goal of classification is to learn the mapping $X \rightarrow Y$, where $y \in Y$ is a class.

Given an item in the sample x_i, where x_i is a vector with one or more variables, the BPM figures out the class label for x_i by performing the following steps:

- Computing the inner product of x_i with a weight vector w.
- Determining whether x_i belongs to a class y, if $(x_i \cdot w)$ is positive. ($x_i \cdot w$ is the inner product of the vectors x_i and w). Gaussian noise is added to the computation.

During training, the BPM algorithm learns the posterior distribution for w using a prior probability distribution and the training data. The Gaussian noise used in the computation helps to address cases where there is no w that can perfectly classify the training data (when the two classes in the training data are not linearly separable).

■ **Note** You can learn more about BPMs at `http://research.microsoft.com/apps/pubs/default.aspx?id=65611`.

In addition, the following tutorial for using Infer.Net and BPMs provides insights into how the algorithm works:

`http://research.microsoft.com/en-us/um/cambridge/projects/infernet/docs/Bayes%20Point%20Machine%20tutorial.aspx`.

Azure Machine Learning provides a Two-Class Bayes Point Machine module, which enables you to build a model that supports binary predictions. The module provides several parameters that can be used to fine-tune the behavior of the Bayes Point Machine module. These include

- **Number of training iterations**: Determines the number of iterations used by the message-passing algorithm in training. Generally, the expectation is that more iterations improve the accuracy of the predictions made by the model. The default number of training iterations is 30.
- **Include bias**: Determines whether a constant bias value is added to each training and prediction instance.

Similar to the Support Vector Machine module, the parameter called `Allow unknown categorical levels` is supported.

Clustering Algorithms

Clustering is a type of unsupervised machine learning. In clustering, the goal is to group similar objects together. Most existing cluster algorithms can be categorized as follows:

- **Partitioning**: Divide a dataset into *k* partitions of data. Each partition corresponds to a cluster.
- **Hierarchical**: Given a dataset, hierarchical clustering approaches start either bottom-up or top-down when constructing the clusters. In the bottom-up approach (also known as agglomerative approach), the algorithm starts with each item in the data set assigned to one cluster. As the algorithm moves up the hierarchy, it merges the individual clusters (that are similar) into bigger clusters. This continues until all the clusters have been merged into one (root of the hierarchy). In the top-down approach (also known as divisive approach), the algorithm starts with all the items in one cluster, and in each iteration, divides into smaller clusters.

- **Density**: Density-based algorithms grow clusters by considering the density (number of items) in the "neighborhood" of each item. They are often used for identifying clusters that have "arbitrary" shapes. In contrast, most partitioning-based algorithms rely on the use of a distance measure. This produces clusters that have regular shapes (such as spherical).

■ **Note** Read a good overview of various clustering algorithms at `http://en.wikipedia.org/wiki/Cluster_analysis`.

In this chapter, we will focus on partitioning-based clustering algorithms. Specifically, you will learn how k-means clustering works.

For partitioning-based cluster algorithms, it is important to be able to measure the distance (or similarity) between points and vectors. Various distance measures include Euclidean, Cosine, Manhattan (also known as City-block) distance, Chebychev, Minkowski, and Mahalanobis distance.

In Azure Machine Learning, the K-Means Clustering module supports Euclidean and Cosine distance measures. Given two points, p1 and p2, the Euclidean distance between p1 and p2 is the length of the line segment that connects the two points. The Euclidean distance can also be used to measure the distance between two vectors. Given two vectors, v1 and v2, the Cosine distance is the cosine of the angle between v1 and v2.

The distance measure used for clustering is chosen based on the type of data being clustered. Euclidean distance is sensitive to the scale/magnitude of the vectors that are compared. For example, even though two vectors seem relatively similar, the scale of the features can affect the value of the Euclidean distance. In this case, the Cosine distance measure is more appropriate, as it is less susceptible to scale. The cosine angle between the two vectors would have been small.

K-means clustering works as follows.

1. Randomly choose k items from the dataset as the initial center for k clusters.
2. For each of the remaining items, assign each one of them to the k clusters based on the distance between the item and the cluster centers.
3. Compute the new center for each of the clusters.
4. Keep repeating steps 2 and 3 until there are no more changes to the clusters, or when the maximum number of iterations is reached.

To illustrate, Figure 6-6 presents a dataset for k-means clustering. There are three distinct clusters in this data, which are illustrated with different colors.

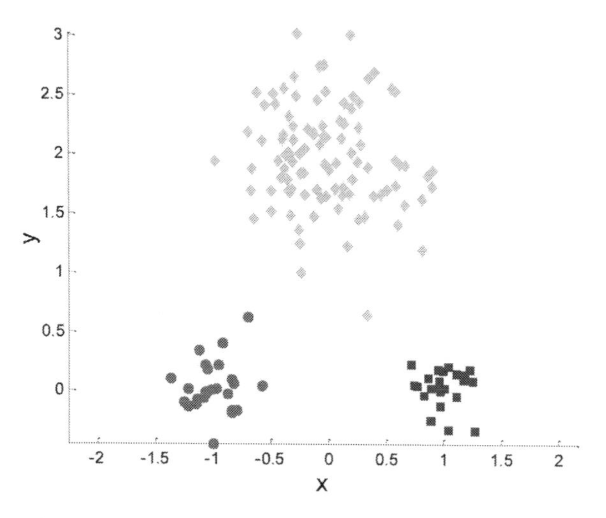

Figure 6-6. *Dataset for k-means clustering*

Figure 6-7 illustrates k-means clustering with k=3 and how the three cluster centroids, represented as + symbols, move each iteration to reduce the mean squared error and more accurately reflect the cluster centers.

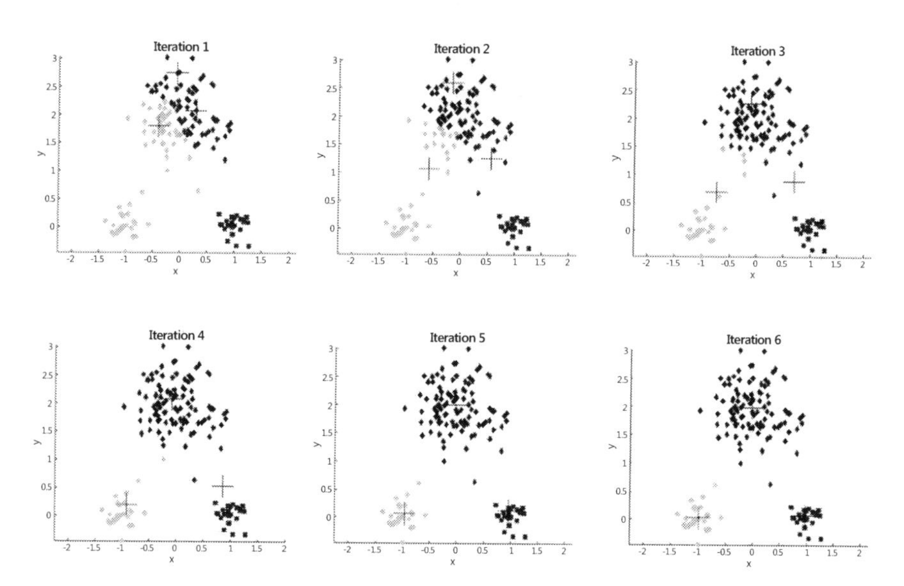

Figure 6-7. *Iterations of the k-means clustering algorithm with k=3 in which the cluster centroids are moving to minimize error*

The K-Means Clustering module in Azure Machine Learning supports a different centroid initialization algorithm. This is specified by the `Initialization` property. Five centroid initialization algorithms are supported. Table 6-1 shows the different centroid initialization methods supported.

Table 6-1. *K-Means Cluster - Centroid Initialization Algorithms*

Centroid Initialization Algorithm	Description
Default	Picks first N points as initial centroids
Random	Picks initial centroids randomly
K-Means++	K-Means++ centroid initialization
K-Means+ Fast	K-Means++ centroid initialization with P:=1 (where the farthest centroid is picked in each iteration of the algorithm)
Evenly	Picks N points evenly as initial centroids

Summary

In this chapter, you learned about different regression, classification, and clustering algorithms. You learned how each of the algorithms work, and the type of problems for which they are suited. The goal of this chapter is to provide you with the foundation for using these algorithms to solve the various problems covered in upcoming chapters. In addition, the resources provided in this chapter will help you learn more deeply about state-of-art machine learning and the theory behind some of these algorithms.

PART III

Practical Applications

CHAPTER 7

Building Customer Propensity Models

This chapter provides a practical guide for building machine learning models. It focuses on buyer propensity models, showing how to apply the data science process to this business problem. Through a step-by-step guide, this chapter will explain how to apply key concepts and leverage the capabilities of Microsoft Azure Machine Learning for propensity modeling.

The Business Problem

Imagine that you are a marketing manager of a large bike manufacturer. You have to run a mailing campaign to entice more customers to buy your bikes. You have a limited budget and your management wants you to maximize return on investment (ROI). So the goal of your mailing campaign is to find the best prospective customers who will buy your bikes.

With an unlimited budget the task is easy: you can simply buy lists and mail everyone. However, this brute force approach is wasteful and will yield a limited ROI since it will simply amount to junk mail for most recipients. It is very unlikely that you will meet your goals with this untargeted approach since it will lead to very low response rates.

A better approach is to use predictive analytics to target the best potential customers for your bikes, such as customers who are most likely to buy bikes. This class of predictive analytics is called buyer propensity models or customer targeting models. With this approach, you build models that predict the likelihood that a prospective customer will respond to your mailing campaign.

In this chapter, we will show you how to build this class of models in Azure Machine Learning. With the finished model you will score prospective customers and only send mail to those who are most likely to respond to your campaign, those prospective customers with the highest probability of response. We will also show how you can maximize the ROI on your limited marketing budget.

■ **Note** You will need to have an account on Azure Machine Learning. Refer to Chapter 2 for instructions to set up your new account if you do not have one yet.

This model we will discuss in this chapter is published as the Buyer Propensity Model in the Azure Machine Learning Gallery. You can access the Gallery at `http://gallery.azureml.net/`. We highly recommend downloading this experiment to your workspace in Azure Machine Learning.

As you saw in Chapter 1, the data science process typically follows these five steps.

1. Define the business problem.
2. Data acquisition and preparation.
3. Model development.
4. Model deployment.
5. Monitor model performance.

Having defined the business problem, your next step is to reframe it as an analytics problem. In this case, your job is to build a statistical model to predict the probability that a person will respond to a mailing campaign, using their demographic data as predictors. Let's now explore the rest of the process, namely data acquisition and preparation, model development, and evaluation, in the rest of this chapter.

Data Acquisition and Preparation

To build the buyer propensity models you will need data from several sources:

- Sales transactions data from your organization's data warehouse. This has useful data about your customers including those who bought bikes in the past.
- Data from previous marketing campaigns that shows which customers were targeted and whether they responded or not.
- Purchased consumer lists from third-party vendors augmented by demographic data from Equifax, Experian, or similar providers. This will have prospective customers to target for the campaign.

This experiment uses the Buyer Propensity Model in the Gallery and includes the required dataset. Be sure to download the model from the Gallery at `http://gallery.azureml.net/`. This model comes with the required dataset for the experiment.

Data Analysis

With the data loaded in Azure Machine Learning, the next step is to do pre-processing to prepare the data for modeling. It is always very useful to visualize the data as part of this process. You can visualize the Bike Buyer dataset by selecting `BikeBuyerWithLocation.csv` from the Saved Datasets menu item in the left pane. When you hover over the small circle at the bottom of the dataset module, a menu opens, giving you two options: Download or Visualize. See Figure 7-1 for details. If you choose the Download option, you can save the data to your local machine and open it in Excel to view the data. Alternatively, you can visualize the data in Azure Machine Learning by choosing the Visualize option.

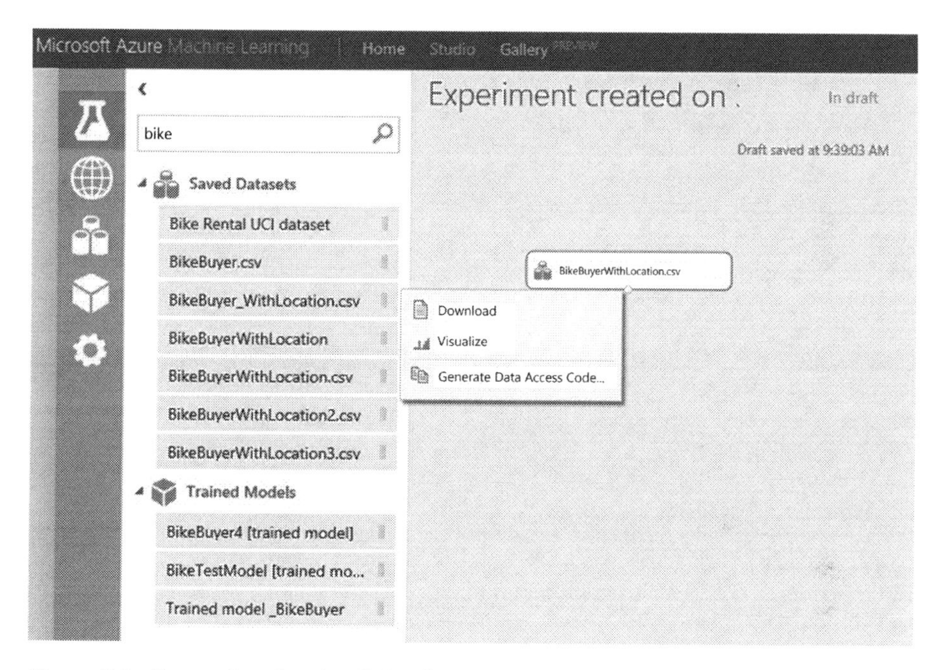

Figure 7-1. *Two options for visualizing data in Azure Machine Learning*

If you choose the Visualize option, the data will be rendered in a new window, as shown in Figure 7-2. Figure 7-2 shows the BikeBuyerWithLocation.csv dataset that has historical sales data on all customers. You can see that this dataset has 10,000 rows and 18 columns, including demographic variables such as marital status, gender, yearly income, number of children, occupation, age, etc. Other variables include home ownership status, number of cars, and commute distance. The last column, called `Bike Buyer`, is very important since it shows which customers bought bikes from your stores in the past.

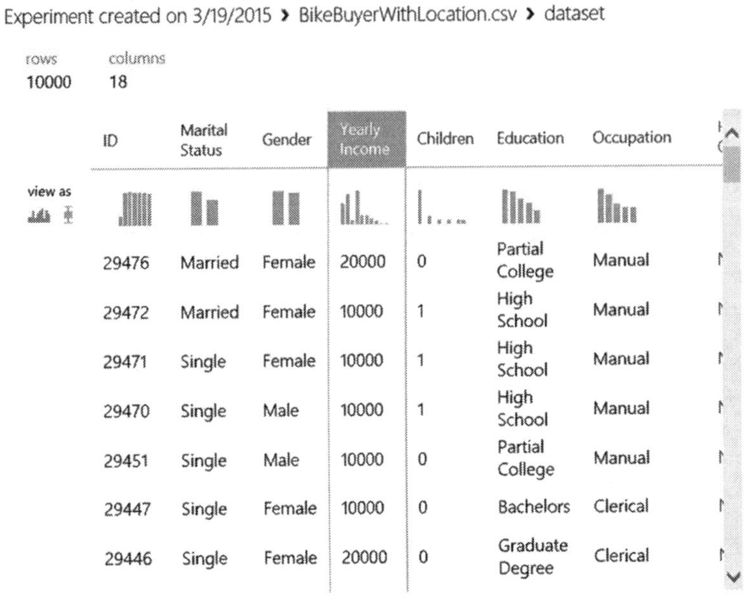

Experiment created on 3/19/2015 › BikeBuyerWithLocation.csv › dataset

rows 10000 columns 18

ID	Marital Status	Gender	Yearly Income	Children	Education	Occupation
29476	Married	Female	20000	0	Partial College	Manual
29472	Married	Female	10000	1	High School	Manual
29471	Single	Female	10000	1	High School	Manual
29470	Single	Male	10000	1	High School	Manual
29451	Single	Male	10000	0	Partial College	Manual
29447	Single	Female	10000	0	Bachelors	Clerical
29446	Single	Female	20000	0	Graduate Degree	Clerical

***Figure 7-2.** Visualizing the Bike Buyer dataset in Azure Machine Learning*

Figure 7-2 also shows the data type of each variable and the number of missing values. By scrolling down you will see a sample of the dataset showing actual values.

The first row at the top of the feature list is a set of thumbnails showing the distribution of each variable. You can see the full distribution by clicking the thumbnail. For instance, if you click the thumbnail above the `Yearly Income` column, Azure Machine Learning opens the full histogram in the window on the top left of the screen; this is shown in Figure 7-3. This histogram shows you the distribution of the selected variable, and you can start to think about the ways in which it can be used in your model. For instance, in Figure 7-3, you can see clearly that yearly income is not drawn from a normal distribution. If anything, it looks more like a lognormal distribution. This is important because some predictive algorithms, such as Linear Regression, assume that the data is normally distributed. To use such algorithms you will have to transform this variable. We will explain later how to transform this yearly income variable to a log scale.

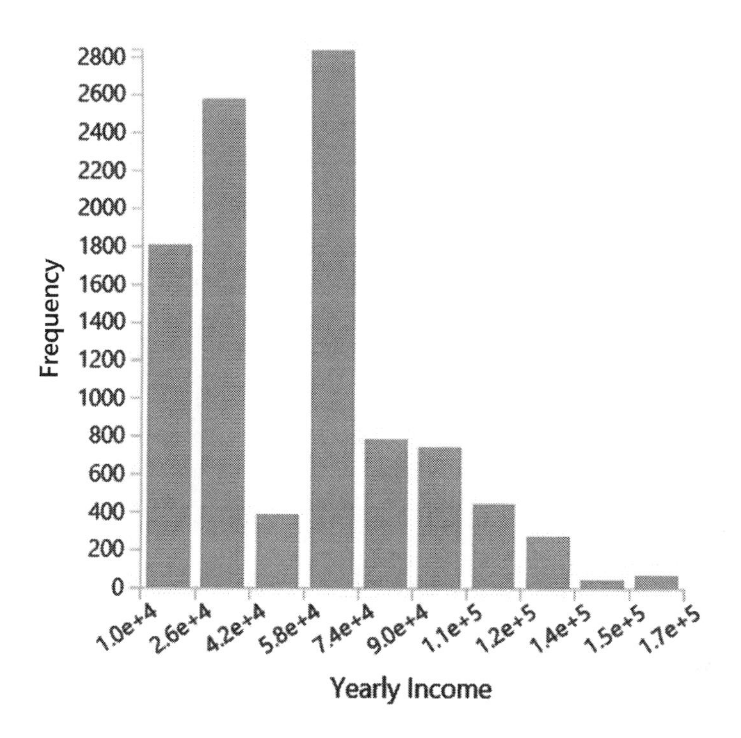

Figure 7-3. *A histogram of yearly income*

Another useful visualization tool in Azure Machine Learning is the box-and-whisker plot that is widely used in statistics. You can visualize your continuous variables with a box-and-whisker plot instead of a histogram. To do this, select the box-and-whisker icon in the first thumbnail labeled **view as**.

Figure 7-4 shows yearly income as a box-and-whisker plot instead of a histogram. On this plot, the y-value at the bottom of the box is the 25th percentile and the value at the top of the box is the 75th percentile. The line inside the box is the median yearly income. The box-and-whisker plot shows outliers much more clearly: in Figure 7-4 the outliers are shown as a single dot above the edge of the whisker. How are outliers determined? A rule of thumb is that any point that lies above or below 1.5*IQR is an outlier. IQR is the interquartile range measured as the 75th percentile - 25th percentile. On the box-and-whisker plot, IQR is the distance between the top and bottom of the box. Since you don't plan to drop the outliers in this case, a good alternative is to transform yearly income with the log function. The result is shown in Figure 7-5. Note that after the log transformation the single dot disappears from the box-and-whisker plot. This means the extreme yearly incomes no longer appear as outliers. This is a great way to include valid extreme values in your model. This treatment can also increase the power of your predictive model.

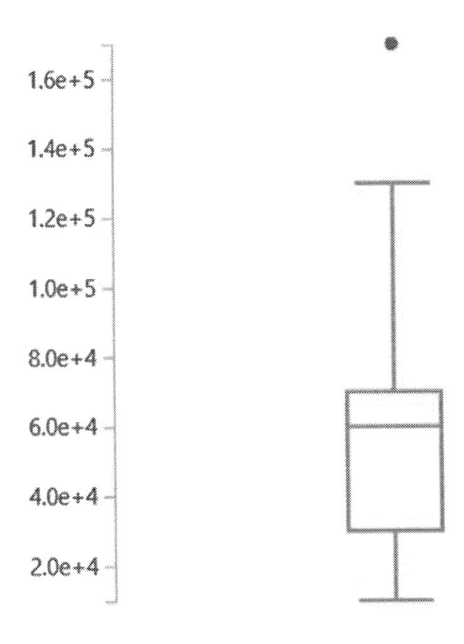

Figure 7-4. *A box-and-whisker plot of yearly income*

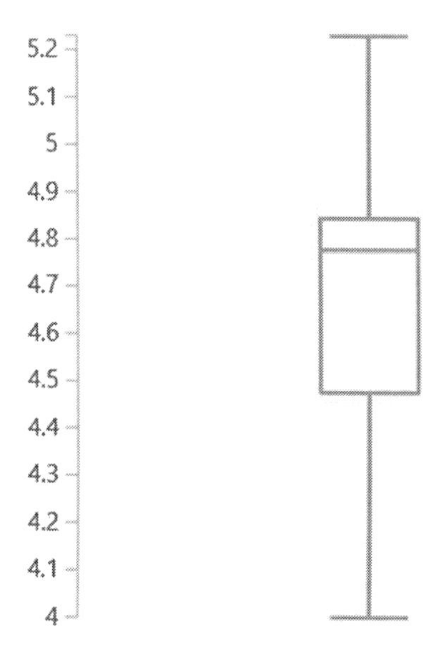

Figure 7-5. *A box-and-whisker plot of the log of yearly income*

How can you transform yearly income to logs? Use the module named **Apply Math Operation** found under the **Statistical Functions menu** in the left pane of Azure Machine Learning. Select **Yearly Income** as the column set and **Log10** from Basic Math Function. This is shown in Figure 7-6.

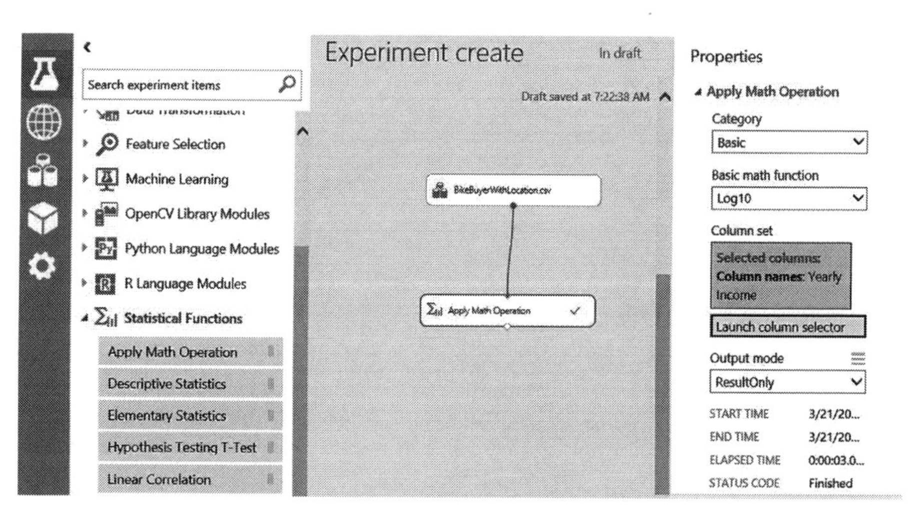

Figure 7-6. *Transforming yearly income to log10 scale*

More Data Treatment

In addition to visualization and data transformation, Azure Machine Learning provides many other options for data preprocessing. The menu on the left pane has many modules organized by function. Many of these modules are useful for data preprocessing.

The Data Format Conversions item has five different modules for converting data formats. For instance, the module named Convert to CSV converts data from different types, such as Dataset, DataTableDotNet, etc., to CSV format.

The Data Transformation item in the menu has subcategories for data filtering, data manipulation, learning with counts, data sampling, and scaling. The menu item named Statistical Functions also has many relevant modules for data preprocessing.

Figure 7-7 shows some of these modules in action. The Join module (found under the Manipulation subcategory) enables you to join datasets. For instance, if there was a second relevant data for the propensity model, you could use the Join module to join it with the Bike Buyer dataset.

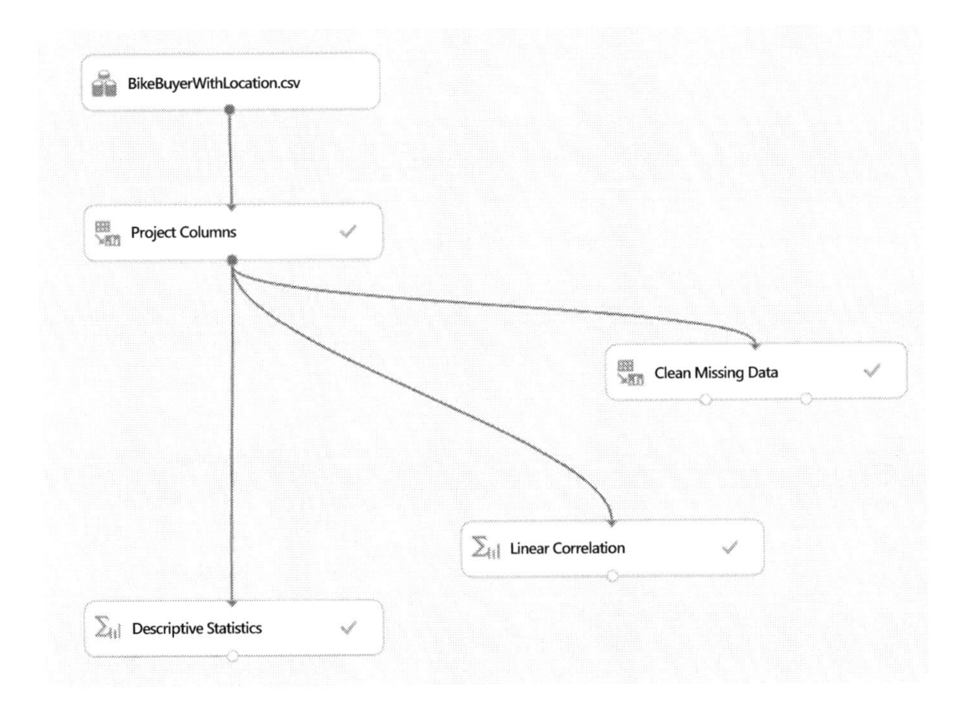

Figure 7-7. *More options for data preparation*

The module named Descriptive Statistics shows descriptive statistics of your dataset. For example, if you click the small circle at the bottom of this module, Azure Machine Learning shows descriptive statistics such as mean, median, standard deviation, skewness kurtosis, etc. of the Bike Buyer dataset. It also shows the number of missing values in each variable.

You can resolve missing values with the module named Clean Missing Data. Like any other module in Azure Machine Learning, when you select this module, its parameters are shown on the right pane. This module allows you to handle missing values in a number of ways. First, you can remove columns with missing values. By default, the tool keeps all variables with missing values. Second, you can replace missing values with a hard-coded value in the parameter box. By default, the tool will replace any missing values with the number 0. Alternatively, you can replace missing values with the mean, median, or mode of the given variable.

The Linear Correlation module is also useful for computing the correlation of variables in your dataset. If you click the small circle at the bottom of the Linear Correlation module, and then select Visualize, the tool displays a correlation matrix. In this matrix, you can see the pairwise correlation of all variables in the dataset. This module calculates the Pearson correlation coefficient. Other correlation types, such as Spearman, are not supported by this module. In addition, note that this module only calculates the correlation for numeric variables. For all non-numeric variables, such as categorical variables, the module shows NaN.

Feature Selection

Feature selection is a very important part of data pre-processing. Also known as variable selection, this is the process of finding the right variables to use in the predictive model. It is particularly critical when dealing with large datasets involving hundreds of variables. Throwing too many variables at a predictive model increases the risk of over-fitting where the model memorizes the data but cannot generalize when tested with unseen data. With too many variables it is also harder to build explainable models. In addition, too many variables increases the chance of collinearity, which is when two or more variables are inter-correlated. Through feature selection you can find the most influential variables for the prediction. Since the Bike Buyer dataset only has 18 variables, you could skip this step. However, let's see how to do feature selection in Azure Machine Learning because it is very important for large datasets.

To do feature selection in Azure Machine Learning, drag the module named **Filter Based Feature Selection** from the list of modules in the left pane. You can find this module by searching for it in the search box or by opening the **Feature Selection** category. To use this module, you need to connect it to a dataset as the input. Figure 7-8 shows it in use as a feature selection for the Bike Buyer dataset. Before running the experiment, use the **Launch column selector** in the right pane to define the target variable for prediction. In this case, choose the column **Bike Buyer** as the target since this is what you have to predict.

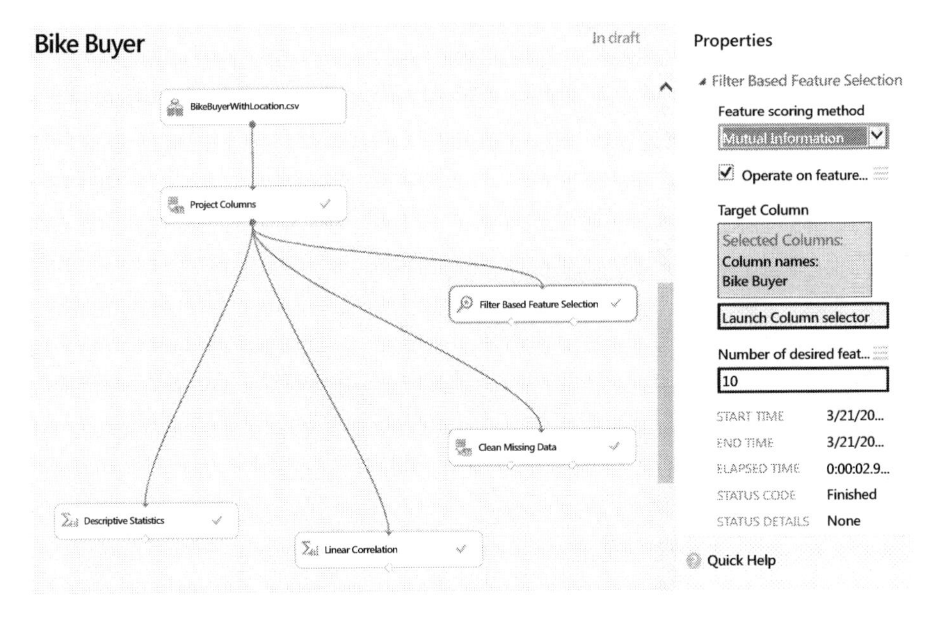

Figure 7-8. *Feature selection in Azure Machine Learning*

You also need to choose the scoring method that will be used for feature selection. Azure Machine Learning offers the following options for scoring:

- Pearson correlation
- Mutual information
- Kendall correlation
- Spearman correlation
- Chi-Squared
- Fischer score
- Count-based

The correlation methods find the set of variables that are highly correlated with the output, but also have low correlation among them. The correlation is calculated using Pearson, Kendall, or Spearman correlation coefficients, depending on the option you choose.

The Fisher score uses the Fisher criterion from statistics to rank variables. In contrast, the mutual information option is an information theoretic approach that uses mutual information to rank variables. Mutual information measures the dependence between the probability density of each variable and that of the outcome variable.

Finally, the Chi-Squared option selects the best features using a test for independence; in other words, it tests whether each variable is independent of the outcome variable. It then ranks the variables based on the results of the Chi-Squared test.

■ **Note** See `http://en.wikipedia.org/wiki/Feature_selection#Correlation_feature_selection` or `http://jmlr.org/papers/volume3/guyon03a/guyon03a.pdf` for more information on feature selection strategies.

When you run the experiment, the **Filter-Based Feature Selection** module produces two outputs. First, the filtered dataset lists the actual data for the most important variables. Second, the module shows a list of the variables by importance with the scores for each selected variable. Figure 7-9 shows the results of the features. In this case, you set the number of features to five and used Chi-Squared for scoring. Figure 7-9 shows six columns since the results set includes the target variable (`Bike Buyer`) plus the top five variables (`ZIP Code`, `City`, `Age`, `Cars`, and `Commute Distance`). The last row of the results shows the score for each selected variable. Since the variables are ranked, the scores decrease from left to right.

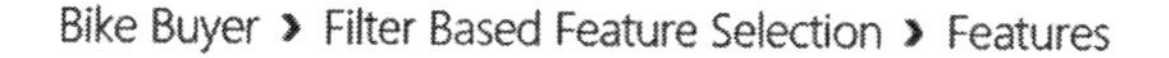

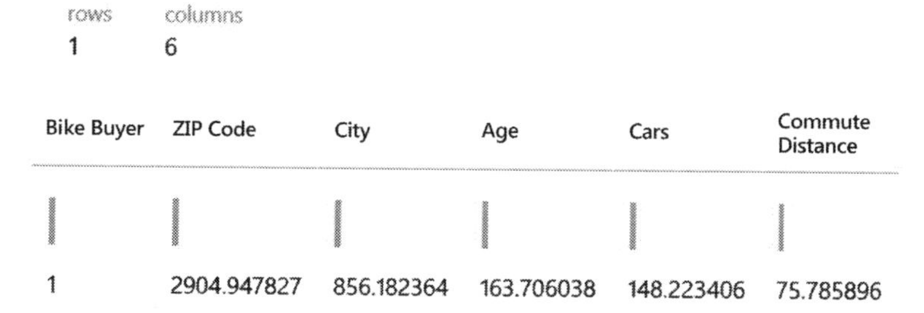

Figure 7-9. The results of feature selection for the Bike Buyer dataset with the top variables

Note that the selected variables will vary based on the scoring method. So it is worth experimenting with different scoring methods before choosing the final set of variables. The Chi-Squared and Mutual Information scoring methods produced a similar ranking of variables for the BikeBuyerWithLocation dataset.

Training the Model

Once the data pre-processing is done, you are ready to train the predictive model. The first step is to choose the right type of algorithm for the problem at hand. For a propensity model, you will need a classification algorithm since your target variable is Boolean. The goal of your model is to predict whether a prospective customer will buy your bikes or not. Hence, it is a great example of a binary classification problem. As you saw in Chapters 1 and 6, classification algorithms use supervised learning for training. The training data has known outcomes for each feature set; in other words, each row in the historical data has the `Bike Buyer` field that shows whether the customer bought a bike or not in the past. The `Bike Buyer` field is the dependent variable (also known as the response variable) that you have to predict. The input variables are the predictors (also known as independent variables or regressors) such as age, cars, commuting distance, education, region, etc. If you use logistic regression, your model can be represented as

$$Bike\ Buyer = \frac{1}{1+e^{-(\beta_0 + \beta_1 Age + \beta_2 Cars + \beta_3 Communte\_distance + \ldots + \varepsilon)}}$$

where β_0 is a constant which is the intercept of the regression line; β_1, β_2, β_3, etc. are the coefficients of the independent variables; and ε is the error that represents the variability in the data that is not explained by the selected variables. In this equation, `Bike Buyer` is the probability that a customer will buy a bike, given his or her input data. Its value ranges from 0 to 1. `Age`, `cars`, and `Communte_distance` are the predictor variables.

During training, you present both the input and the output variables to the selected algorithm. At each learning iteration, the algorithm will try to predict the known outcome using the current weights that have been learned so far. Initially, the prediction error is high because the weights have not been learned adequately. In subsequent iterations, the algorithm will adjust its weights to reduce the predictive error to the minimum. Note that each algorithm uses a different strategy to adjust the weights such that predictive error can be reduced. See Chapter 6 for more details on the statistical and machine learning algorithms in Azure Machine Learning.

Azure Machine Learning offers a wide range of classification algorithms from multiclass decision forest and jungle to two-class logistic regression, neural network, and support vector machines. For a customer propensity model, a two-class classification algorithm is appropriate since the response variable (`Bike Buyer`) has two classes. So you can use any of the two-class algorithms under the Classification subcategory. To see the full list of algorithms, expand the category named **Machine Learning** in the left pane. Select **Initialize Model** from this menu. Expand the subcategory named **Classification** and Azure Machine Learning will list all available classification algorithms. We recommend experimenting with a few of these algorithms until you find the best one for the job. Figure 7-10 shows a simple but complete experiment for the bike buyer propensity model.

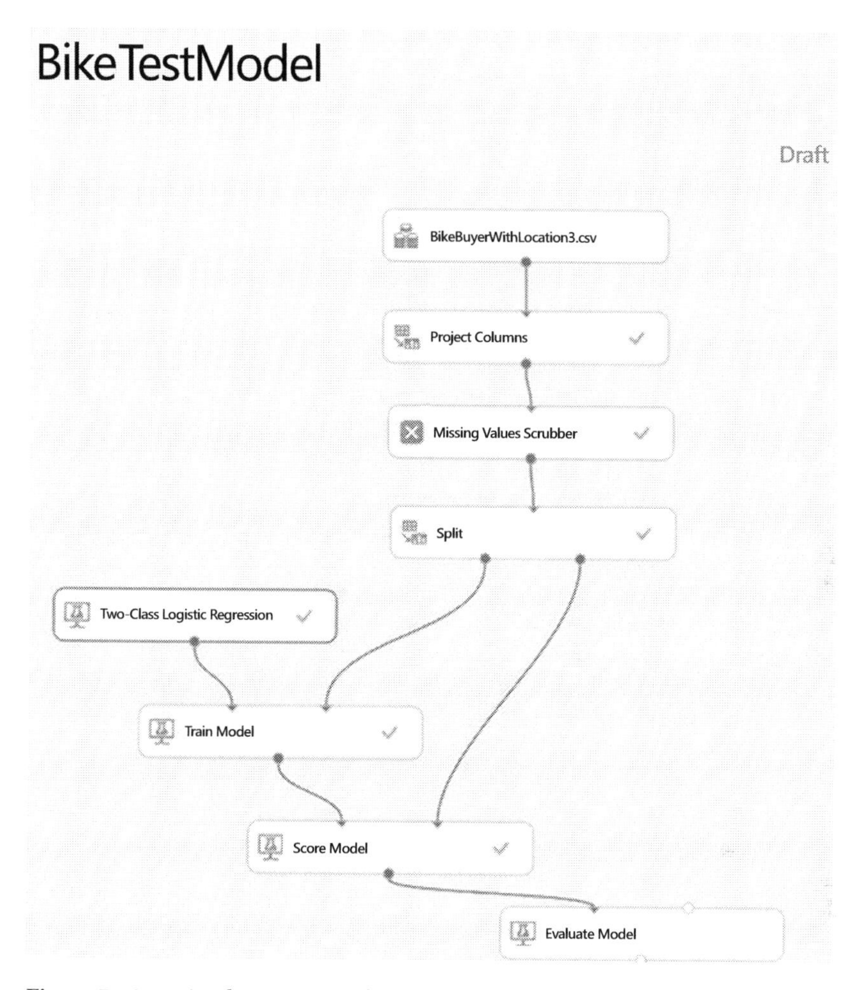

***Figure 7-10.** A simple experiment for customer propensity modeling*

The Project Columns module simply excludes the columns named `ID`, `latitude`, `longitude`, and `country` since they are not relevant to the model. Use the Clean Missing Data module to handle missing values, as discussed in the previous section. The Split module splits the data into two samples, one for training and the second for testing. In this experiment, we reserved 70% of the data for training and the remaining 30% for testing. This is one strategy commonly used to avoid over-fitting. Without the test sample, the model can easily memorize the data and noise. In that case, it will show very high accuracy for the training data but will perform poorly when tested with unseen data in production. Another good strategy to avoid over-fitting is cross-validation; this will be discussed later in this chapter.

Two modules are used for training the predictive model: the Two-Class Logistic Regression module implements the logistic regression algorithm, while the Train Model actually trains the learning algorithm. The Train Model module trains any suitable algorithm to which it is connected. So it can be used to train any of the classification modules discussed earlier, such as Two-Class Boosted Decision Tree, Two-Class Decision Forest, Two-Class Decision Jungle, Two-Class Neural Network, or Two-Class Support Vector Machine.

Model Testing and Validation

After the model is trained, the next step is to test it with a hold-out sample to avoid over-fitting. In this example, your test set is the 30% sample you created earlier with the Split module. Figure 7-10 shows how to use the module named Score Model to test the trained model.

Finally, the module named Evaluate Model is used to evaluate the performance of the model. Figure 7-10 shows how to use this module to evaluate the model's performance on the test sample. The next section also provides more details on model evaluation.

As mentioned earlier, a strategy to avoid over-fitting is cross-validation where you use not just two, but multiple samples of the data to train and test the model. Azure Machine Learning uses 10-fold validation where the original dataset is split into 10 samples. Figure 7-11 shows a modified experiment that uses cross-validation as well as the train and test set approach. On the right track of the experiment you replace the two modules, **Train Model** and **Score Model**, with the **Cross Validate Model** module. You can check the results of the cross-validation by clicking the **small circle** on the bottom right side of the **Cross Validate Model** module. This shows the performance of each of the 10 models.

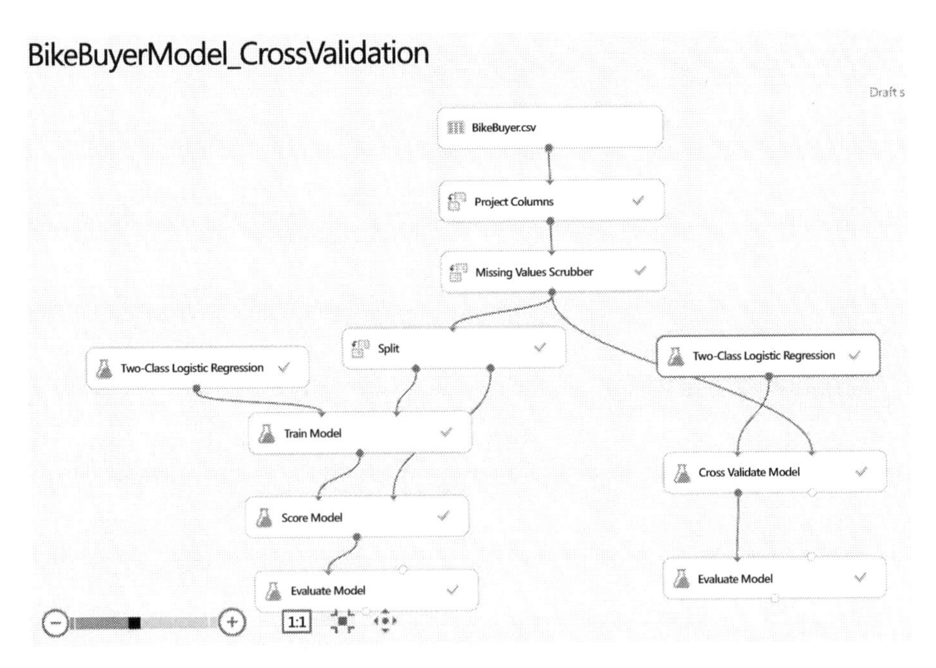

Figure 7-11. *A modified experiment with cross-validation*

Model Performance

The Evaluate Model module is used to measure the performance of a trained model. This module takes two datasets as inputs. The first is a scored dataset from a tested model. The second is an optional dataset for comparison.

After running the experiment you can check your model's performance by clicking the **small circle** at the bottom of the module **Evaluate Model**. This module provides the following metrics to measure the performance of a classification model such as the propensity model:

- The **Receiver Operating Curve** (ROC) curve plots the rate of true positives to false positives.
- The **Lift curve** (also known as the Gains curve) plots the number of true positives vs. the positive rate. This is popular in marketing.
- The **Precision versus recall** chart.
- The **Confusion matrix** shows type I and II errors.

Figure 7-12 shows the ROC curve for the propensity model you built earlier. The ROC curve visually shows the performance of a predictive binary classification model. The diagonal line from (0,0) to (1,1) on the chart shows the performance of random guessing; so if you randomly selected who to target, your response would be on this diagonal line. A good predictive model should do much better than random guesses. Hence, on the ROC curve, a good model should fall above the diagonal line. The ideal model that is 100% accurate will have a vertical line from (0,0) to (0,1), followed by a horizontal line from (0,1) to (1,1).

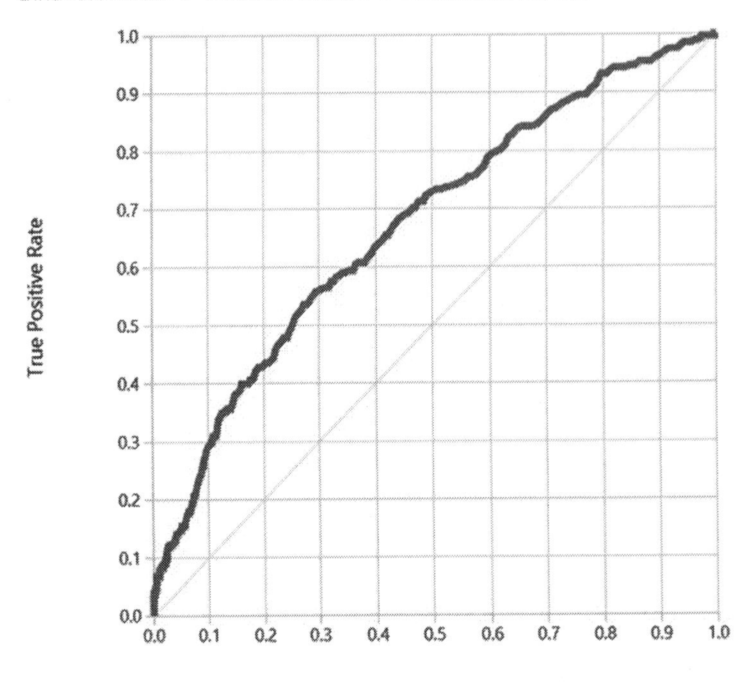

Figure 7-12. *The ROC curve for the customer propensity model*

One way to measure the performance from the ROC curve is to measure the area under the curve (AUC). The higher the area under the curve, the better the model's performance. The ideal model will have an AUC of 1.0, while a random guess will have an AUC of 0.5. The logistic regression model you built has an AUC of 0.67, which is much better than a random guess!

Figure 7-13 shows the confusion matrix for the logistic regression model you built earlier. The confusion matrix has four cells, namely

- **True positives**: These are cases where the customer actually bought a bike and the model correctly predicts this.
- **True negatives**: In the historical dataset these customers did not buy bikes, and the model correctly predicts that they would not buy.
- **False positives**: In this case, the model incorrectly predicts that the customer would buy a bike when in fact they did not. This is commonly referred to as a Type I error. The logistic regression you built had no false positives.

- **False negatives**: Here the model incorrectly predicts that the customer would not buy a bike when in real life the customer did buy one. This is also known as a Type II error. The logistic regression model had up to 270 false negatives.

Figure 7-13. *Confusion matrix and more performance metrics*

It is worth noting that when we test the model, its output is the predicted probabilities for each example. So we need to set a threshold probability to determine the predicted class. By default, Azure Machine Learning uses a threshold of 0.5. Hence if the predicted probability is greater than 0.5, the predicted class is set to 1, and zero otherwise. However, 0.5 in this case is arbitrary. A better approach is to set the probability threshold at the point where the sensitivity equals the specificity. Sensitivity and specificity are discussed in detail at the end of this chapter.

In addition, Figure 7-13 also shows the accuracy, precision, and recall of the model. Here are the formulas for these metrics.

Precision is the rate of true positives in the results.

$$Precision = \frac{tp}{tp + fp} = \frac{8}{8 + 0} = 1.0$$

Recall is the percentage of buyers that the model identifies and is measured as

$$Recall = \frac{tp}{tp + fn} = \frac{8}{8 + 270} = 0.029$$

Finally, accuracy measures how well the model correctly identifies buyers and non-buyers, as in

$$Accuracy = \frac{tp + tn}{tp + tn + fp + fn} = \frac{8 + 2722}{8 + 2722 + 0 + 270} = 0.91$$

where tp = true positive, tn = true negative, fp = false positive, and fn = false negative.

You can also compare the performance of two models on a single ROC chart. As an example, let's modify the experiment in Figure 7-11 to use the **Two-Class Boosted Decision tree** module as the second trainer instead of the **Two-Class Logistic Regression** module. Now your experiment has two different classifiers, the Two-Class Logistic Regression on the left branch and the Two-Class Boosted Decision Tree on the right branch. You connect the scored datasets from both models into the same **Evaluate Model** module. The updated experiment is shown in Figure 7-14 and the results are illustrated in Figure 7-15 and Figure 7-16. On this chart, the curve labeled **scored dataset to compare** is the ROC curve for the **Two-Class Boosted Decision Tree** model, while the one labeled **scored dataset** is the one for the **Two-Class Logistic Regression**. You can see clearly that the boosted decision tree model outperforms the logistic regression model, as it has a higher lift over the diagonal line for random guesses. The area under the curve (AUC) for the boosted decision tree model is 0.849, which is better than that of the logistic regression model, which was 0.67, as you saw earlier.

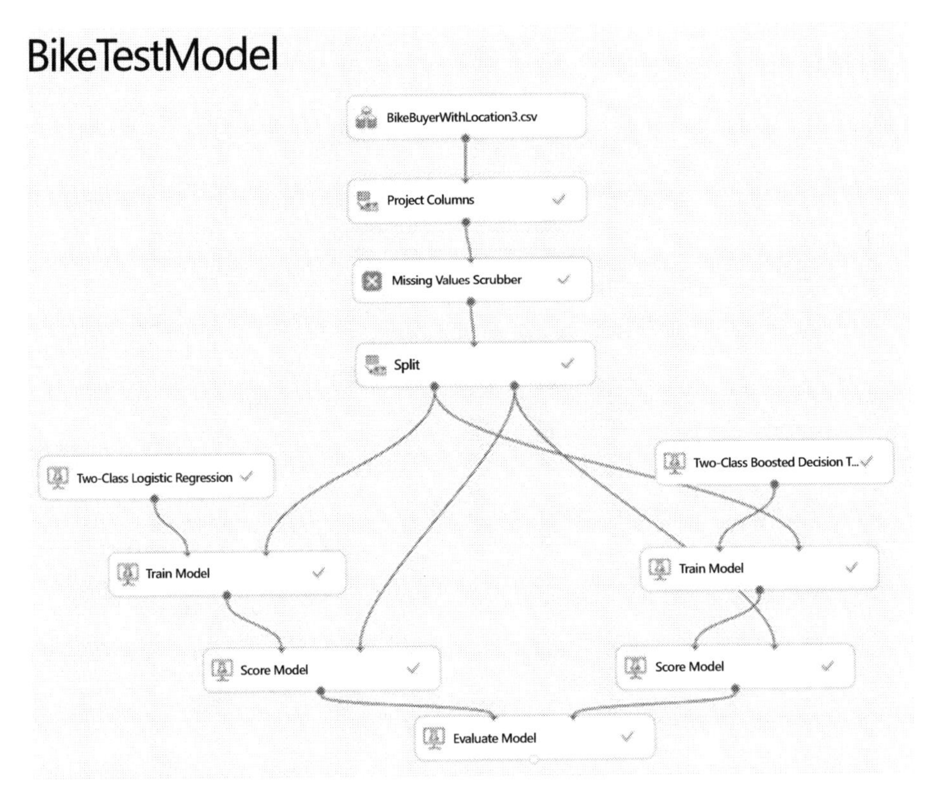

Figure 7-14. *Updated experiment that compares two algorithms, Logistic Regression and Boosted Decision Trees*

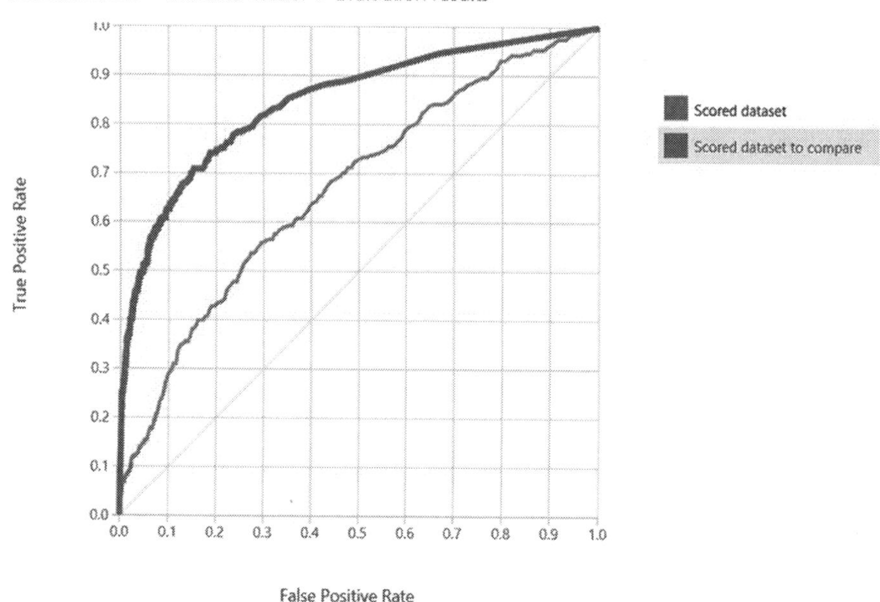

Figure 7-15. *ROC curves comparing the performance of two predictive models, the boosted decision tree vs. logistic regression*

True Positive	False Negative	Accuracy	Precision	Threshold	AUC
128	150	0.919	0.579	0.5	0.849
False Positive	**True Negative**	**Recall**	**F1 Score**		
93	2629	0.460	0.513		

Figure 7-16. *Results of the Boosted Decision Tree model*

Prioritizing Evaluation Metrics

It is clear that the Boosted Decision Tree model is better than that of Logistic Regression because the Boosted Decision Tree model has higher area under the curve (AUC) and accuracy than the Logistic Regression model. So it makes sense to use the Boosted Decision Tree model for this customer targeting exercise. For each model you have seen that Azure Machine Learning provides several evaluation metrics such as accuracy, precision, recall, and the F1 score. Which of these is the most important? It is tempting to focus on accuracy as the most important variable. After all, the accuracy measures the model's ability to correctly classify test cases. In general, the higher the accuracy, the lower the error rates. So it seems intuitive to prioritize this metric for assessing the performance of our models.

However, accuracy alone can be misleading. For most buyer propensity models, accuracy is not the right performance to prioritize because of class imbalance. In this example, the BikeBuyerWithLocation dataset has only 1,000 customers who bought bikes, out of a total of 10,000 customers. So a naïve model that predicts 0 for all prospects will have an accuracy of 90%, even though it would incorrectly classify all the customers who actually bought bikes. This is due to the prevalence of one class in the dataset since the majority of customers did not buy bikes.

To avoid this trap, you need to check the *sensitivity* and *specificity* of your model before deploying it in production. Sensitivity measures the model's ability to correctly identify the target of interest. In this case, sensitivity measures how well a model will predict those who bought bikes. Sensitivity can be measured as follows:

$$Sensitivity = \frac{number\ of\ true\ positives}{number\ of\ true\ positives + number\ of\ false\ negatives} = \frac{tp}{tp + fn} = Recall$$

In contrast, specificity measures a model's ability to correctly identify negative cases. In this example, it measures how well a model will predict which customers will not buy bikes. Specificity is given by

$$Specificity = \frac{number\ of\ true\ negatives}{number\ of\ true\ negatives + number\ of\ false\ positives}$$

$$= \frac{number\ of\ true\ negatives}{number\ of\ customers\ who\ did\ not\ buy\ bikes}$$

$$= \frac{tn}{tn + fp}$$

The ideal model would have 100% sensitivity (it would perfectly predict who will buy bikes) and 100% specificity (it would perfectly predict who will not buy bikes). However, this is not possible because each model has a theoretical error limit, also known as the Bayes Error rate, which rules out this possibility. So for most practical purposes we have to make a tradeoff between sensitivity and specificity. This is why the F1 score is important. The F1 score is the harmonic mean of sensitivity and specificity. A good model that balances sensitivity and specificity will have a higher F1 score. The F1 score is given by

$$F1\,Score = \frac{2 * tp}{2 * tp + fp + fn}$$

From the results of the two models, you see that the Boosted Decision Tree model has an F1 score of 0.513 while that of the Logistic Regression model is only 0.056, despite a perfect precision of 1.0! This is yet another reason why the Boosted Decision Tree model is better than the one based on Logistic Regression.

One final factor to consider is explainability. If your stakeholders simply want the most accurate model and are not too concerned about how it works, then the Boosted Decision Tree model is better. However, if your stakeholders want to explain predictions from the model, then a Boosted Decision Tree may not be as desirable because its results are harder to explain. Since it is an ensemble model, its prediction for each customer comes from several trees. So it is harder to explain what variables contribute to the prediction for a single customer. This makes the Boosted Decision Tree a black box. So if the stakeholder needs to explain the results of each prediction, the Logistic Regression model is better suited since you can obtain the important factors from the variable weights (the coefficients β_1, β_2, β_3, etc.) in the Logistic Regression equation. The weights have to be interpreted with care especially if the scale of some of the variables is much higher than that of the rest. In general, the weights are easier to interpret if the variables are either binary or normalized. In addition, you need to check the p-values to find the variables that are statistically significant. Typically you may reject any variables whose p-values are greater than 5%.

Summary

In this chapter, we provided a practical guide on how to build buyer propensity models with the Microsoft Azure Machine Learning service. You learned how to perform data pre-processing and analysis, which are critical steps towards understanding the data that you are using to build customer propensity models. With that understanding of the data, you used a two-class logistic regression and a two-class boosted decision tree algorithm to perform classification. You also saw how to evaluate the performance of your models and determine the best one for your application. Finally, we reviewed key performance metrics such as accuracy, precision, recall (also known as sensitivity), specificity, and the F1 score. In this section, you saw why it is naïve to rely on accuracy for classification, and why the F1 score is more appropriate for a buyer propensity model.

CHAPTER 8

Visualizing Your Models with Power BI

Building predictive models is essential. However, explaining the results is just as important. Even an excellent predictive model can be seriously undermined by a failure to effectively communicate the results. Data visualization helps data scientists explain the results of predictive models to their stakeholders and end users. In this chapter, we show how you can share the results of your models through Power BI.

Overview

There is a large body of literature on data visualization that is beyond the scope of this book. This chapter focuses on Microsoft's Power BI. You will learn how to use it to share the results of your model through visualization. You will explore three approaches for visualizing your results with Power BI.

1. You'll score a test dataset in Azure Machine Learning and use Power BI tools in Microsoft Excel for visualization.
2. In the second approach, you will learn how to score your test dataset in Excel by calling a trained predictive model through the REST API provided by Azure Machine Learning, when you publish an experiment as a web service. This enables you to score and visualize data without leaving Excel.
3. In the third approach, you will score your test data in Azure Machine Learning and visualize it with Microsoft's Power BI service at `www.powerbi.com`. This service enables you to visualize your results without installing Excel 2013 on your machine.

Introducing Power BI

This section presents a brief introduction to Power BI. We provide the information you need to visualize the results of your predictive models in Power BI. For more in-depth coverage of Power BI, please see the resources listed at the end of this section.

Power BI provides exciting new ways to visualize data, share results, and collaborate in new ways. The new experience of Power BI is based on powerbi.com, an online service that you can use to create and share reports and dashboards seamlessly. Both the old and new Power BI experiences are deeply integrated with Excel, which is Microsoft's leading tool for business analysts. The new Power BI service at powerbi.com was released into general availability on July 24th. Let's review the key components of Power BI.

The focal point of the new powerbi.com is a dashboard that allows you to visualize all your data in one place. Figure 8-1 shows an example of the dashboard that ships with powerbi.com. The live dashboard allows you to load data from several sources including your Excel workbooks. You can also visualize data from SQL Server Analysis (SSAS) on premises. In addition, powerbi.com ships with out-of-the box connectors for cloud Software as a Service (SaaS) solutions such as Salesforce, Zendesk, Marketo, SendGrid, GitHub, Dynamics CRM Online, and Dynamics AX. With these connectors you can visualize your data from any of these products through dashboards in powerbi.com. The dashboard supports several chart types, from the standard bar charts and pie charts to combo charts, maps, gauges, and new funnel charts.

Figure 8-1. A sampe dashboard in powerbi.com

Microsoft ships Power BI Designer, a desktop version of Power BI that you can use to create your reports and dashboards in an offline mode. You can connect to the same data sources as the Power BI service to load data and create reports and dashboards. When you are done, you can share these with others by publishing to the Power BI Service. This allows you to do personal BI in a safe environment before sharing broadly online. For those without Excel 2013 this is a good way to create reports quickly and cheaply. So if you just need to build a BI dashboard that loads data from sources other than Excel, the Power BI Designer is a great choice. However, if you already own a license for Excel 2013 or have a lot of your data in Excel, then you can simply build your data models in Excel and share the results as dashboards in PowerBI.com.

The best authoring experience on premises is in Excel, which is Microsoft's leading tool for business analysis. Excel has rich BI capabilities such as Power Query, PowerPivot, Power View, and Power Map. You can create rich BI models on your desktop and then share these by publishing to the Power BI service (powerbi.com). Let's review the key BI features in Excel 2013:

- **Power Query** is an Excel add-in that you can use to discover, combine, and refine data from multiple sources of relational and non-relational data. Power Query allows you to pull data from traditional databases such as SQL Server, Oracle, Teradata, IBM DB2, and non-relational sources including Facebook, Hadoop (both HDFS and Microsoft Azure HDInsight), SharePoint lists, Wikipedia, etc. This tool significantly simplifies the process of loading and transforming data from these myriad sources and analyzing it in Excel. You can download the add-in from the Microsoft Download site at `www.microsoft.com/en-us/download/details.aspx?id=39379`. More details on Power Query are available at `https://support.office.com/en-us/article/Introduction-to-Microsoft-Power-Query-for-Excel-6E92E2F4-2079-4E1F-BAD5-89F6269CD605`.
- **PowerPivot** is an in-memory BI engine that ships natively in Excel 2013. Before 2013, PowerPivot was available as an Excel add-in. However, in Excel 2013, PowerPivot ships natively and is now called the Data Model. With PowerPivot you can create a power BI model that uses data from multiple sources all in Excel. You can add any data you load, including unstructured data, with the Power Query add-in described above. The Data Model in Excel supports the DAX language for intuitive calculations, has a Diagram View for managing relationships in your data, allows you to define calculated fields and key performance indicators, and many more features. Due to a more efficient storage model, the Data Model allows you to load very large datasets into Excel. With enough memory you can load over one billion rows of data into Excel, thanks to this in-memory column store. You are only limited by your computer's memory.

- **Power View** is an interactive tool for exploring and visualizing your data. You can use it to build dynamic and ad hoc reports easily in Excel even with very little BI experience. Power View was an Excel add-in before Excel 2013. However, it now ships natively in Excel 2013. Power View works in tandem with PowerPivot: when you create a data model in memory with PowerPivot, you can use Power View to visualize the data through rich interactive reports. Power View is also a feature of SharePoint 2013.
- **Power Map** is a very cool 3D mapping tool that you can use to visualize geospatial data. For example, you can use it to map sales across different countries or cities with very little effort. The key is to have location markers in your data. And the location data does not have to be longitudes and latitudes only. Instead, Power Map can map data using any relevant address attributes such as country, city, ZIP code, etc. Power Map is an add-in for Excel 2013. You can download it from `www.microsoft.com/en-us/download/details.aspx?id=38395`. More information on Power Map is also available at `www.microsoft.com/en-us/powerBI/power-map.aspx`.

■ **Note** Please refer to `http://powerbi.com` and `www.youtube.com/user/mspowerbi` for more details on Power BI.

Three Approaches for Visualizing with Power BI

Now let's explore the three approaches for visualizing your data with Power BI. Your first visualization approach will be to score a test dataset in Azure Machine Learning Studio and then visualize it with Power BI tools in Excel.

■ **Note** In this chapter, you will use the Bike Buyer model from Chapter 2. This model is published as the Buyer Propensity Model in the Azure Machine Learning Gallery. You can access the Gallery at `http://gallery.azureml.net/`.

Download this experiment to your workspace in Azure Machine Learning for the rest of the experiment. Figure 8-2 shows the Buyer Propensity model in Azure Machine Learning Studio.

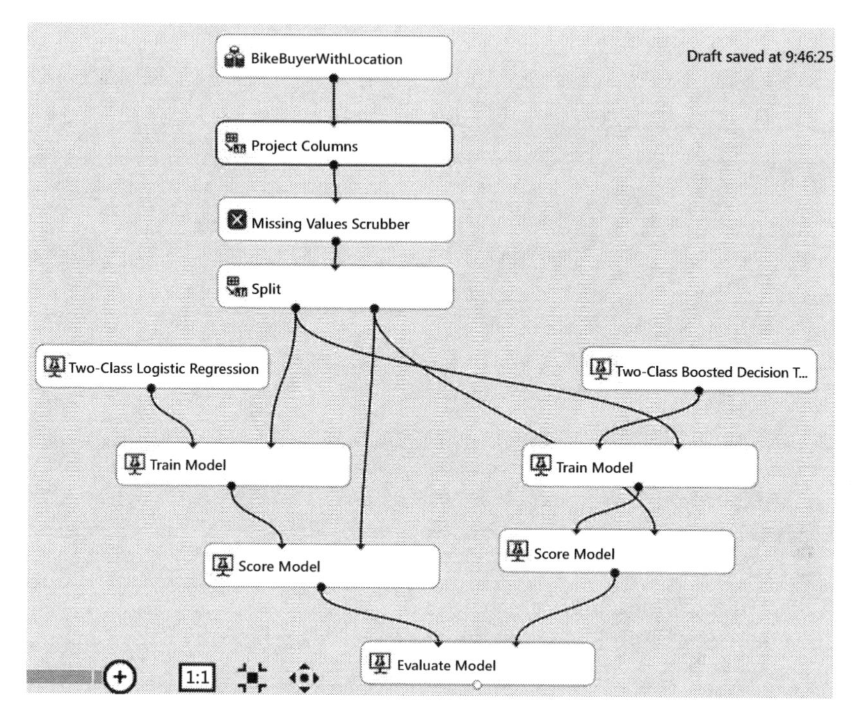

Figure 8-2. *The Buyer Propensity Model in Azure Machine Learning Studio*

Scoring Your Data in Azure Machine Learning and Visualizing in Excel

To start with, you will modify the Buyer Propensity Model to include geospatial fields such as latitude, longitude, address, city, and country. These fields were excluded from the model since they were not statistically relevant. Now you are adding them back because you will need them to visualize the results on a map.

Modify the Buyer Propensity Model to include geospatial fields with the following steps.

1. Go to the **Buyer Propensity Model** in Azure Machine Learning Studio.
2. Drag down the dataset box named **BikeBuyerWithLocation** to the bottom right side of the canvas.
3. In the *first* **Project Columns** module at the top of the experiment, use the **Launch Column Selector** to add the variable **ID**. To do this,
    a. Click the **Project Columns** module, and in the **right pane**, click **Launch column selector**. You will see the excluded variables in the text box, as shown in Figure 8-3.

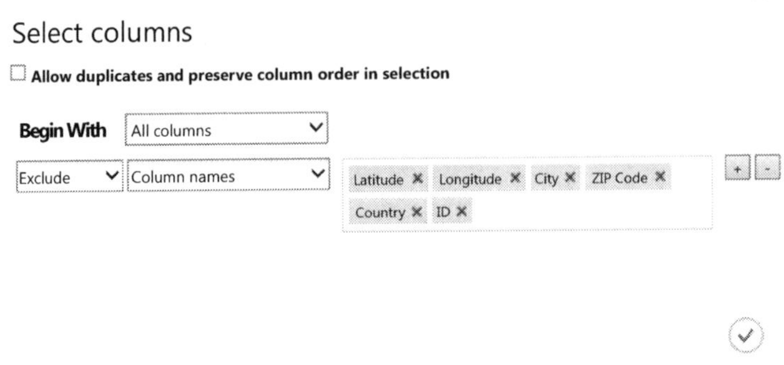

Select columns
Allow duplicates and preserve column order in selection
Begin With
All columns
Exclude
Column names
Latitude
Longitude
City
ZIP Code
Country
ID

Figure 8-3. Output of the first Project Columns module showing the list of excluded variables

   b. Drop the variable ID from the list of excluded variables by clicking the **X** on the ID variable in this list.

4. Add the following three new modules to the bottom of your experiment: **Project Columns, Join**, and **Convert to CSV**. Connect them as shown in the diagram in Figure 8-4

Two-Class Boosted Decision T...
BikeBuyerWithLocation
Train Model
Project Columns
Score Model
Join
Convert to CSV

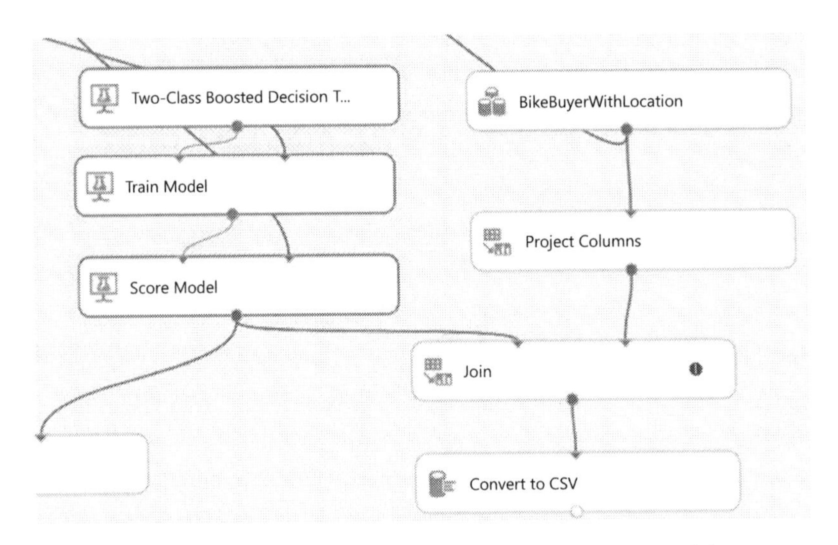

Figure 8-4. The Buyer Propensity Model modified to add geospatial data

5. In the new **Project Columns** module, add the variables **ID**, **Longitude**, **Longitude**, **City**, **ZIP Code**, and **Country**. See Figure 8-5 for details.

Figure 8-5. *Adding ID and geospatial variables to the Project Columns module*

6. Next, in the **Join** module, use the variable **ID** to join your two datasets as follows:

   a. In the properties pane, use **Launch column selector** under **Join key columns for L**. Set the value to **ID**.

   b. Use **Launch column selector** under **Join key columns for R**. Also, set its value to **ID**. See Figure 8-6 for the parameters in the **Join** module.

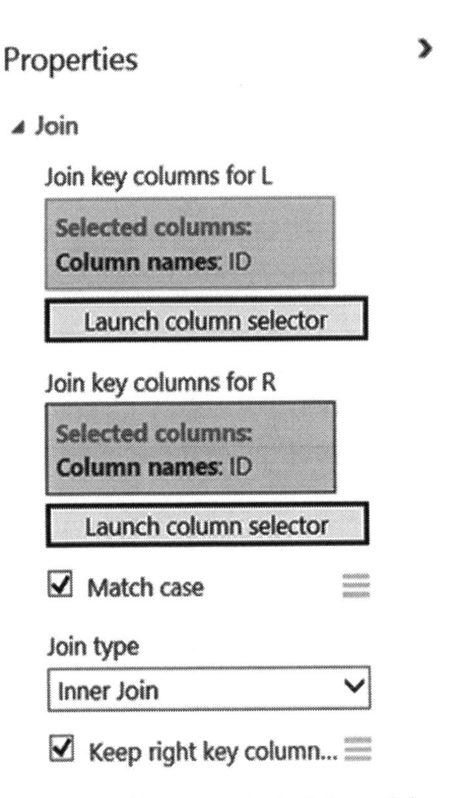

Figure 8-6. *Parameters in the Join module*

7. Run the experiment. On completion, select the small dot on the **Convert to CSV** module, right-click it, and choose **download** from the menu.

Now that you have a scored dataset that contains a geospatial variable, you need to map your data. Next, let's switch to Excel where you will visualize your scored dataset. To this, follow these instructions.

1. Open the resulting CSV dataset in Excel.
2. Save the file as an Excel workbook and name it **BikeBuyerwithLocation2_Scored_dataset.xlsx**.
3. Select Power Map from the **Insert** ribbon in Excel.
4. Click **Launch Power Map**.
5. In the right pane, choose **City** in the pane named **Geography and Map Level**, and click **Next**.
6. Then drag the field called **Scored Labels** to the **Height** box. This plots the scored labels on the map.

7. Also, drag the field **Scored Labels** to the **Category** box. This will use the two categories (Yes and No) in the scored labels. See Figure 8-7 for the parameters in Power Map.

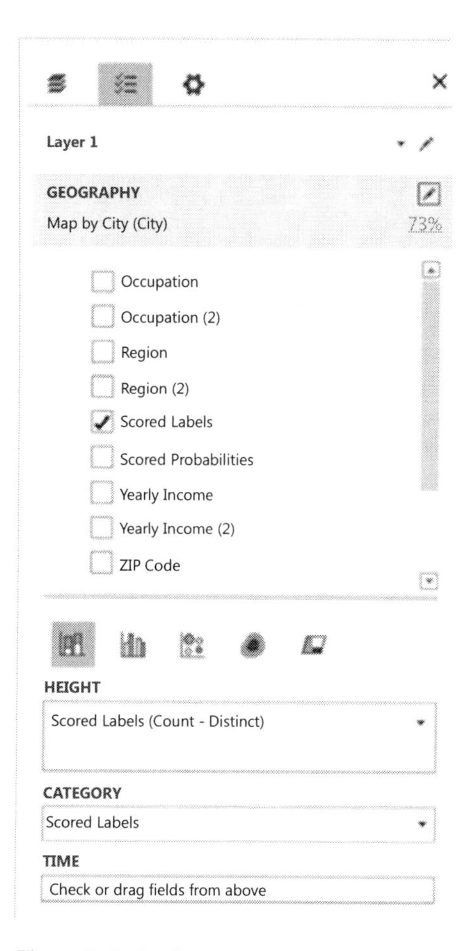

Figure 8-7. *Setting parameters in Power Map*

8. Click **Map Labels** in the ribbon to label the map.

9. Click **2D Map** from the ribbon, and plot Top 100 locations by **YES**. This shows a ranking of top cities that are most likely to respond to the campaign.

Congratulations! You have just completed your first visualization in Power BI. Your map should appear as shown in Figure 8-8. This map plots customers' propensity to buy by city. At each city, you can see the number of customers predicted to buy (shown in blue bars) or not buy (shown in orange bars). The overlaid 2D chart shows a list of cities sorted by their propensity to buy (sorted by Yes). You can see that the top cities by propensity are West Jordan (Utah), Spokane (Washington), and Nampa (Idaho).

Figure 8-8. *A 3D plot of the Scored Labels in Power Map*

Scoring and Visualizing Your Data in Excel

The second way to visualize your results is to score your model and visualize the results in Excel. In Chapter 2, you saw how to publish your experiment to run as a web service in Azure Machine Learning. Follow the steps in Chapter 2 to publish the Buyer Propensity Model as a web service in Azure Machine Learning. Name your experiment **BikeTestModelScore**. When you publish your model, Azure Machine Learning will automatically create an Excel spreadsheet containing an API key that you will need to access your mode. Now follow these steps to access the Excel spreadsheet containing code for calling the REST API.

1. In Azure Machine Learning, click **Web Services** on the left pane to see your web services. One of them should be the **BikeTestModelScore** that you published.

2. Click **BikeTestModelScore**. This opens a new page showing full details of your published service. Figure 8-9 shows an illustration of what you should see. The API key for your new web service is shown in the text box named API key. You will need this key to securely access your web service. There are also two URLs listed here: the first is the URL you will need to call the service interactively to score one row of data at a time. The second URL can be used to score multiple rows in batch mode with a single API call. Also listed on this page is a link to an Excel workbook from where you can also score your data.

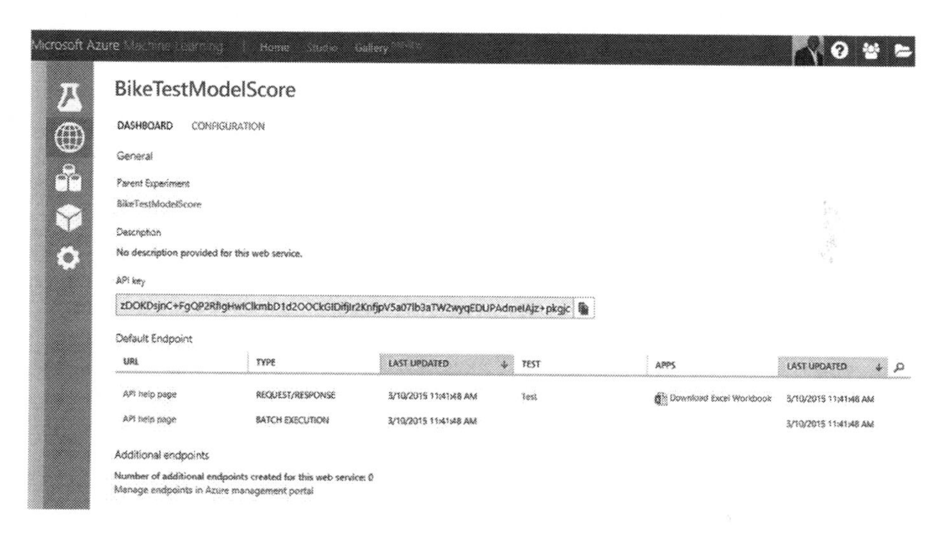

Figure 8-9. Details of the published Buyer Propensity Model in Azure Machine Learning

3. Click the Excel link and it will download the file. Open the file in Excel. Once you enable macros, Excel will show your input parameters and the model's output from the web service. You can enter values for the input parameters. Once you are done, Excel will automatically compute the outputs by calling the API of your web service. Figure 8-10 shows an example. In the last section, you downloaded a CSV file containing test data. You can also copy and paste several rows of input data from this CSV file into the new Excel spreadsheet. Excel will automatically compute the output of each row.

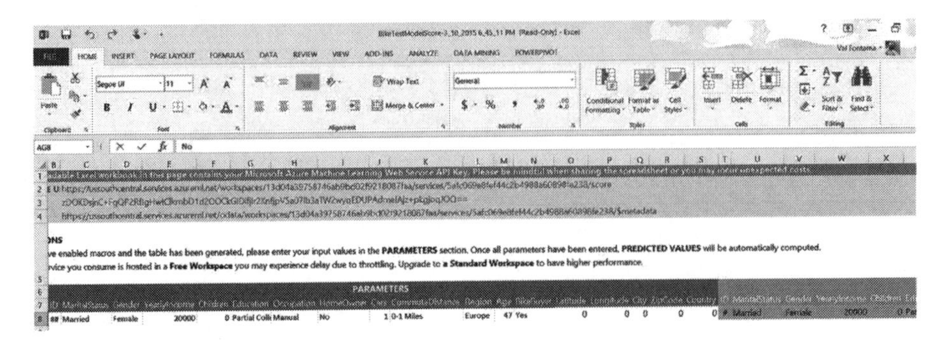

Figure 8-10. *Testing a model from Excel*

Scoring Your Data in Azure Machine Learning and Visualizing in powerbi.com

The third approach to visualizing your results is to use powerbi.com. As you saw earlier, powerbi.com is an online service that you can use to visualize your data that resides on premises or in the cloud. The centerpiece of powerbi.com is the dashboard. The dashboard enables you to visualize your data using several chart types in exciting ways. Through its Q&A feature you can also search your data easily using English text. This is great for your users since they do not need to learn SQL to query the data. Also, powerbi.com offers a desktop tool called Power BI Designer that you can use to create your reports and dashboards offline. When you are ready, you can publish it online at powerbi.com. Power BI is also available as an iOS app that runs on iPads and iPhones.

Let's get started. To use powerbi.com you need to sign up at `www.powerbi.com/`, which is now available as a free preview service. When you first sign up, you will see the Retail dashboard shown in Figure 8-1. Your goal is to create your own dashboards and reports using the results of the Buyer Propensity Model.

To create your own dashboard in powerbi.com you have to load your results dataset. Let's learn how to do this.

Loading Data

In the first approach to visualization you saved a scored dataset in the file named **BikeBuyerwithLocation2_Scored_dataset.xlsx**. You will now load this file from your local filesystem using the following steps.

1. Click **Get Data** from the top left pane. You will see that you can load data from several sources including Excel, Power BI Designer (the desktop version of Power BI), SQL Server Analysis Services, Microsoft Dynamics CRM, and non-Microsoft products such as Marketo, Salesforce, and GitHub. This is shown in Figure 8-11.

Figure 8-11. *Showing the data sources for powerbi.com*

2. To load your scored dataset, select **Excel Workbook** and then click **Connect**. Choose **Computer** and then **Browse**. Select the file named **BikeBuyerwithLocation2_Scored_dataset.xlsx** from your computer. When the data is loaded, you will see it under the **Datasets** menu in the left pane. Now you are ready to start building your report.

Building Your Dashboard

When your dataset is loaded, you will see a new dashboard named **BikeBuyerwithLocation2_Scored_dataset.xlsx** under the Dashboard menu on the left pane. This is a blank canvas you will use to create your new dashboard. Figure 8-12 shows this blank canvas. Now create your own dashboard with this dataset through the following steps.

Figure 8-12. *A blank canvas ready for your first dashboard*

1. On the right pane is a filter. The filter shows all the fields available to create your charts. Click the field named **Country**. By default, Power BI will plot a 2D map of the sum of IDs by country. You should choose the right variable to visualize by country.

2. Next, you might want to see how many prospective customers will respond to the mailing campaign by country. To do that, replace **ID** with the **predicted scores**. After you have replaced the ID, unselect **ID** in the filters list, and instead select **Scored Labels**. Now the 2D map plots Scored Labels by country, so it shows how many prospective customers will or will not buy bikes by country.

3. Next, add a second chart to show how the propensity to buy bikes varies with the level of education. To do this, follow these steps.

    a. Click the field named **Education** in the filters. By default, Power BI will plot a second chart showing the sum of IDs by Education. Again, this is not very informative. So you need to change the Y-axis to a more useful variable than the sum of IDs.

    b. Unselect the **sum of IDs** field in the **Filters** pane, and instead drag the field named **Scored Labels** to the **values** section of the **Filters** pane. See Figure 8-13 for details.

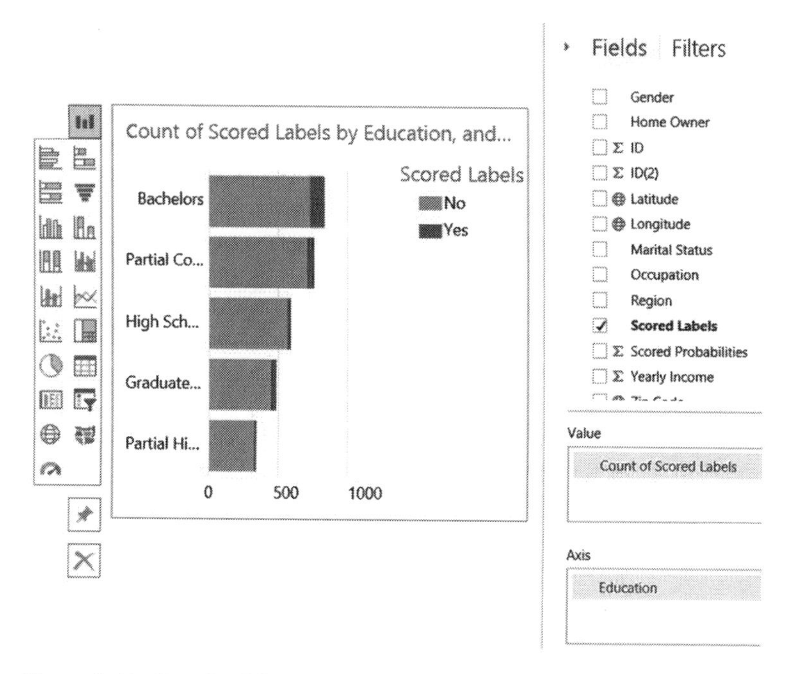

Figure 8-13. *Details of the second chart that plots the count of Scored Labels by Education*

c. Finally, convert the chart type to a horizontal stacked bar as follows:

- Select the new bar chart of **Scored Labels by Education**.
- Choose the top small icon on the top left of this chart. It is called the **change visualization type**.
- Now select the **stacked bar** type. The result is shown in Figure 8-13.

At the end of these steps, your dashboard should appear as shown in Figure 8-14. This dashboard has two charts: the first is a 2D map showing propensity to buy by country, while the second shows propensity to buy by level of education.

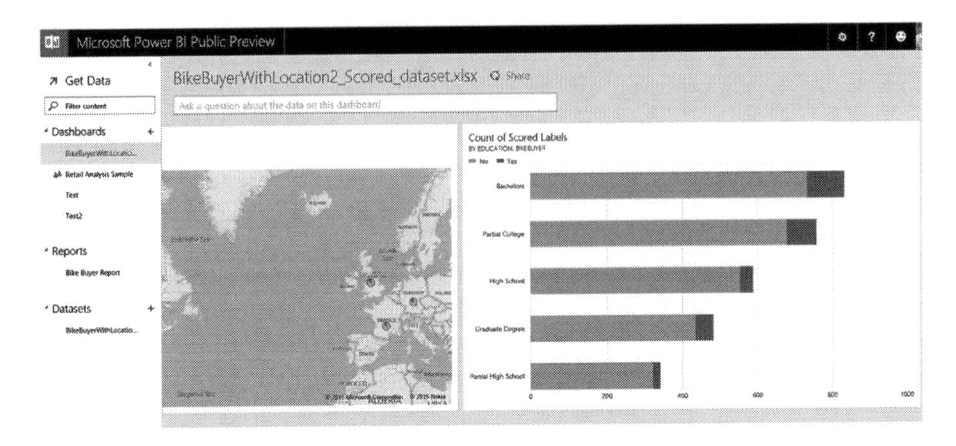

Figure 8-14. *Complete dashboard with two charts*

Summary

Data visualization is a critical tool for a data scientist because it helps to communicate the results of modeling to stakeholders. Even after creating an excellent predictive model, you need to communicate the insights obtained from the model to your stakeholders. In this chapter, you learned how to visualize the results of your models with Power BI. This chapter started with an introduction to Power BI from Microsoft. You gained essential skills to help you visualize your results with Power BI in three ways.

1. You scored a test dataset in Azure Machine Learning and used Power BI tools in Microsoft Excel for visualization.

2. You scored your test dataset in Excel by calling a published predictive model through the REST API of Azure Machine Learning. This enabled you to score and visualize data without leaving Excel.

3. You scored your test data in Azure Machine Learning and visualized it with Microsoft's Power BI service at `www.powerbi.com`.

You are now ready to create rich visualizations of your results with Power BI. With the knowledge you gained from this chapter you can now impress your stakeholders with dazzling dashboards that communicate the findings from your Machine Learning models. Now go forth and impress!

CHAPTER 9

Building Churn Models

In this chapter, we reveal the secrets of building customer churn models, which are in very high demand. Many industries use churn analysis as a means of reducing customer attrition. This chapter will show a holistic view of building customer churn models in Microsoft Azure Machine Learning.

Churn Models in a Nutshell

Businesses need to have an effective strategy for managing customer churn because it costs more to attract new customers than to retain existing ones. Customer churn can take different forms, such as switching to a competitor's service, reducing the time spent using the service, reducing the number of services used, or switching to a lower-cost service. Companies in the retail, media, telecommunication, and banking industries use churn modeling to create better products, services, and experiences that lead to a higher customer retention rate.

Let's drill deeper into why churn modeling matters to telecommunication companies. The consumer business of many telecommunication companies operates in an intensely competitive market. In many countries, it is common to have two or more telecommunication companies competing for the same customer. In addition, mobile number portability makes it easier for customers to switch to another telecommunication provider.

Many telecommunication companies track customer attrition (or customer churn) as part of their annual report. The use of churn models has enabled telecommunication providers to formulate effective business strategies for customer retention, and to prevent potential revenue loss.

Churn models enable companies to predict which customers are most likely to churn, and to understand the factors that cause churn to occur. Among the different machine learning techniques used to build churn models, classification algorithms are commonly used. Azure Machine Learning provides a wide range of classification algorithms including decision forest, decision jungle, logistic regression, neural networks, Bayes Point Machines, and support vector machines. Figure 9-1 shows the different classification algorithms that you can use in Azure Machine Learning Studio.

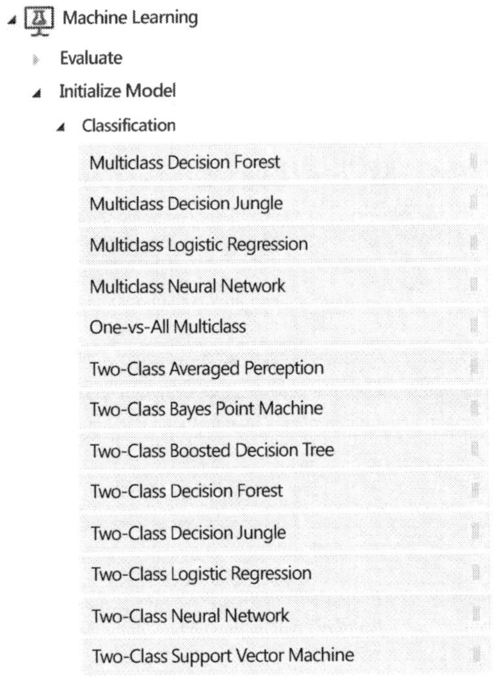

- Machine Learning
  - Evaluate
  - Initialize Model
    - Classification
      - Multiclass Decision Forest
      - Multiclass Decision Jungle
      - Multiclass Logistic Regression
      - Multiclass Neural Network
      - One-vs-All Multiclass
      - Two-Class Averaged Perception
      - Two-Class Bayes Point Machine
      - Two-Class Boosted Decision Tree
      - Two-Class Decision Forest
      - Two-Class Decision Jungle
      - Two-Class Logistic Regression
      - Two-Class Neural Network
      - Two-Class Support Vector Machine

Figure 9-1. *Classification algorithms available in ML Studio*

Prior to building the churn model (based on classification algorithms), understanding the data is very important. Given a dataset that you are using for both training and testing the churn model, you should ask the following questions (non-exhaustive) about the data:

- What kind of information is captured in each column?
- Should you use the information in each column directly, or should you compute derived values from each column that are more meaningful?
- What is the data distribution?
- Are the values in each column numeric or categorical?
- Does a column consists of many missing values?

Once you understand the data, you can start building the churn model using the following steps.

1. Data preparation and understanding
2. Data preprocessing and feature selection
3. Classification model for predicting customer churn

4. Evaluating the performance of the model
5. Operationalizing the model

In this chapter, you will learn how to perform each of these steps to build a churn model for a telecommunication use case. You will learn the different tools that are available in Azure Machine Learning Studio for understanding the data and performing data preprocessing. And you will learn the different performance metrics that are used for evaluating the effectiveness of the model. Let's get started!

Building and Deploying a Customer Churn Model

In this section, you will learn how to build a customer churn model using different classification algorithms. For building the customer churn model, you will be using a telecommunication dataset from KDD Cup 2009. The dataset is provided by a leading French telecommunication company, Orange. Based on the Orange 2013 Annual Report, Orange has 236 million customers globally (15.5 million fixed broadband customers and 178.5 mobile customers).

The goal of the KDD Cup 2009 challenge is to build an effective machine learning model for predicting customer churn, willingness to buy new products/services (appetency), and opportunities for upselling. In this section, you will focus on predicting customer churn.

■ **Note** KDD Cup is an annual competition organized by the ACM Special Interest Group on Knowledge Discovery and Data Mining (SIGKDD). Each year, data scientists participate in various data mining and knowledge discovery challenges. These challenges range from predicting who is most likely to donate to a charity (1997), clickstream analysis for an online retailer (2000), predicting movie rating behavior (2007), to predicting the propensity of customers to switch providers (2009).

Preparing and Understanding Data

In this exercise, you will use the small Orange dataset, which consists of 50,000 rows. Each row has 230 columns (referred to as variables). The first 190 variables are numerical and the last 40 variables are categorical.

Before you start building the experiment, download the following small dataset and the churn labels from the KDD Cup website:

- `orange_small_train.data.zip` from `www.sigkdd.org/sites/default/files/kddcup/site/2009/files/orange_small_train.data.zip`
- `orange_small_train_churn.labels` from `www.sigkdd.org/sites/default/files/kddcup/site/2009/files/orange_small_train_churn.labels`

In the `orange_small_train_churn.labels` file, each line consists of a +1 or -1 value. The +1 value refers to a positive example (the customer churned), and the -1 value refers to a negative example (the customer did not churn).

Once the file has been uploaded, you should upload the dataset and the labels to Machine Learning Studio as per these steps.

1. Click **New** and choose **Dataset ➤ From Local File** (Figure 9-2).

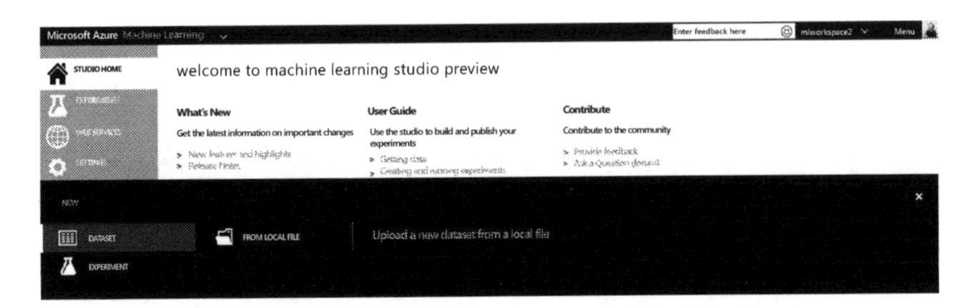

***Figure 9-2.** Uploading the Orange dataset using Machine Learning Studio*

2. Next, choose the file to upload: **orange_small_train.data** (Figure 9-3).

Upload a new dataset

Select the data to upload:
D:\book\CloudML\Customer Churn\orar Browse...

This is the new version of an existing dataset

Existing dataset:
orange_small_train.data

Select a type for the new dataset:
Generic TSV File with a header (.tsv)

Provide an optional description:
Customer Churn dataset

***Figure 9-3.** Uploading the dataset*

3. Click **OK**.

After the Orange dataset has been uploaded, repeat the steps to upload the churn labels file to Machine Learning Studio. Once this is done, you should be able to see the two Orange datasets when you create a new experiment. To do this, create a new experiment, and expand the Saved Datasets menu in the left pane. Figure 9-4 shows the Orange training and churn labels datasets that you uploaded.

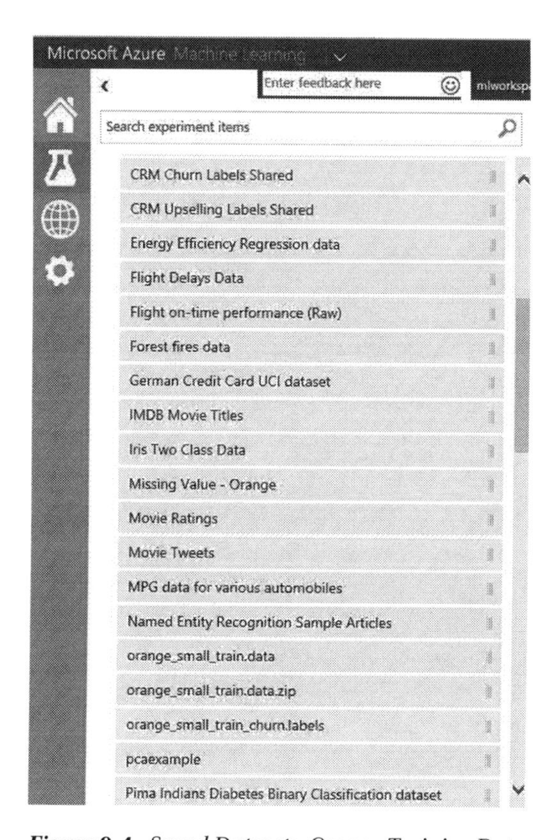

Figure 9-4. *Saved Datasets, Orange Training Data and Churn Labels*

When building any machine learning model, it is very important to understand the data before trying to build the model. To do this, create a new experiment as follows.

1. Click **New ➤ Experiment**.
2. Name the experiment **Understanding the Orange Dataset**.
3. From **Saved Datasets**, choose the **orange_small_train.data dataset** (double-click it).
4. From **Statistical Functions**, choose **Descriptive Statistics** (double-click it).

5. You will see both modules. Connect the dataset with the **Descriptive Statistics** module. Figure 9-5 shows the completed experiment.

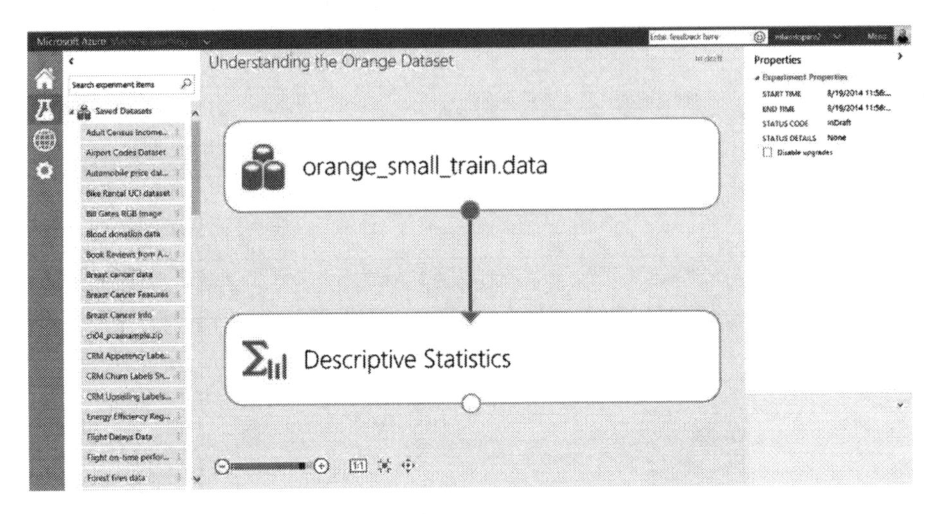

Figure 9-5. *Understanding the Orange dataset*

6. Click **Run**.

7. Once the run completes successfully, right-click the circle below **Descriptive Statistics** and choose **Visualize**.

8. You will see an analysis of the data, which covers Unique Value Count, Missing Value Count, Min, and Max for each of the variables (Figure 9-6).

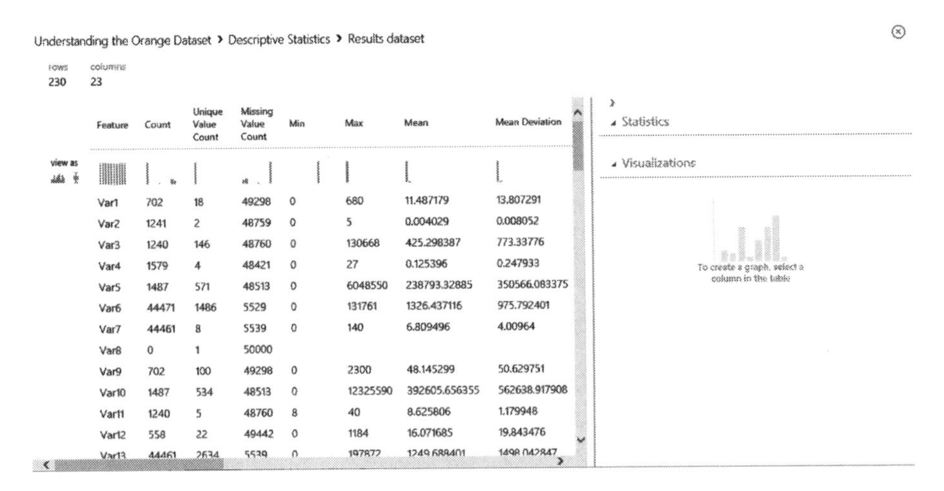

Understanding the Orange Dataset › Descriptive Statistics › Results dataset

rows 230 columns 23

Feature	Count	Unique Value Count	Missing Value Count	Min	Max	Mean	Mean Deviation
Var1	702	18	49298	0	680	11.487179	13.807291
Var2	1241	2	48759	0	5	0.004029	0.008052
Var3	1240	146	48760	0	130668	425.298387	773.33776
Var4	1579	4	48421	0	27	0.125396	0.247933
Var5	1487	571	48513	0	6048550	238793.32885	350566.083375
Var6	44471	1486	5529	0	131761	1326.437116	975.792401
Var7	44461	8	5539	0	140	6.809496	4.00964
Var8	0	1	50000				
Var9	702	100	49298	0	2300	48.145299	50.629751
Var10	1487	534	48513	0	12325590	392605.656355	562638.917908
Var11	1240	5	48760	8	40	8.625806	1.179948
Var12	558	22	49442	0	1184	16.071685	19.843476
Var13	44461	2634	5539	0	197872	1249.688401	1498.042847

Figure 9-6. *Descriptive statistics for the Orange dataset*

This provides useful information on each of the variables. From the visualization, you will observe that there are lots of variables with missing values (such as `Var1`, `Var8`). For example, `Var8` is a row with almost no useful information.

■ **Tip** When visualizing the output of **Descriptive Statistics**, it shows the top 100 variables. To see all of the statistics for all 230 variables, right-click the bottom circle of the **Descriptive Statistic** module and choose **Save as dataset**. After the dataset has been saved, you can choose to download the file and see all the rows in Excel.

Data Preprocessing and Feature Selection

In most classification tasks, you will often have to identify which of the variables should be used to build the model. Machine Learning Studio provides two feature selection modules that can be used to determine the right variables for modeling. This includes filter-based feature selection and linear discriminant analysis.

For this exercise, you will not be using these feature selection modules. You learned how to use these feature selection modules in Chapter 8. Figure 9-7 shows the data preprocessing steps.

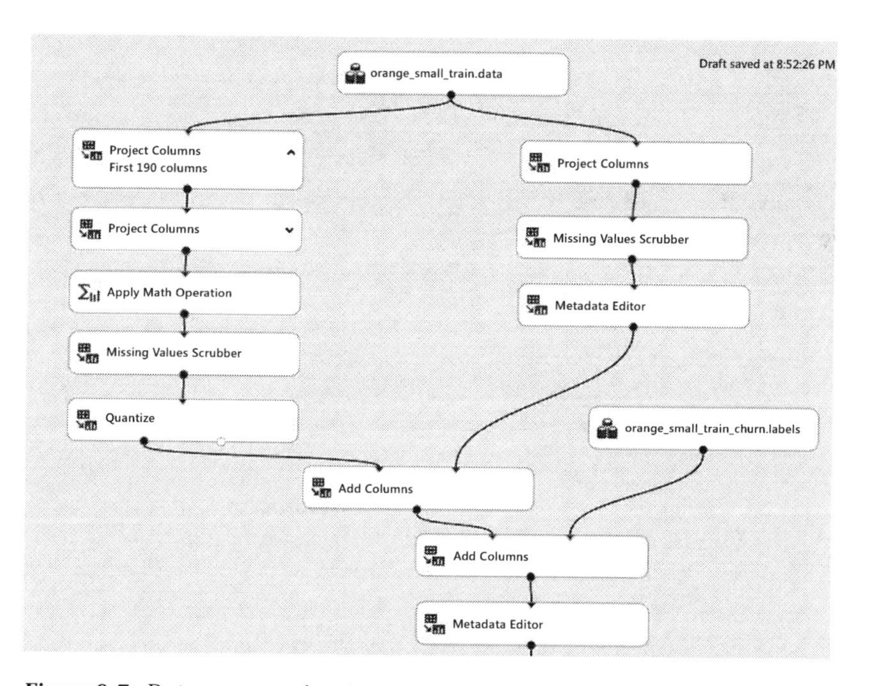

Figure 9-7. Data preprocessing steps

For simplicity, perform the following steps to preprocess the data.

1. Divide the variables into the first 190 columns (numerical data) and the remaining 40 columns (categorical data). To do this, add two **Project Columns** modules.

   For the first **Project Column** module, select **Column indices: 1-190** (Figure 9-8).

   For the second **Project Column** module, select **Column indices: 191-230** (Figure 9-9).

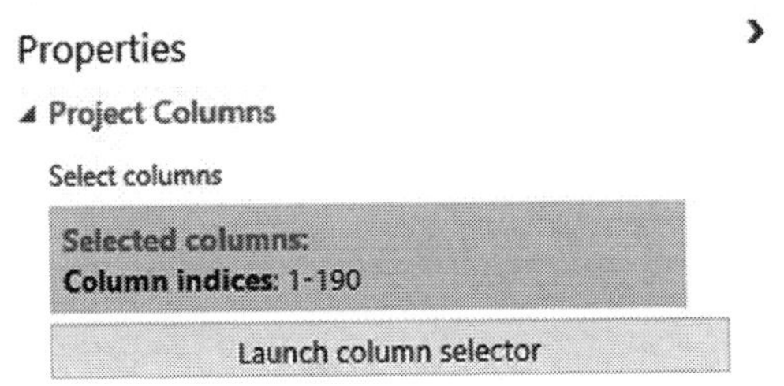

Properties

Project Columns

Select columns

Selected columns:
Column indices: 1-190

Launch column selector

Figure 9-8. *Selecting column indices 1-190 (numerical columns)*

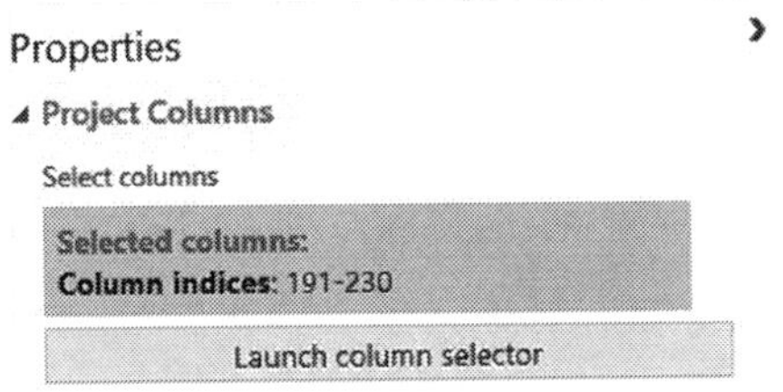

Properties

Project Columns

Select columns

Selected columns:
Column indices: 191-230

Launch column selector

Figure 9-9. *Selecting column indices 191-230 (categorical columns)*

2. For the first 190 columns, do the following.

   a. Use **Project Columns** to select the columns that contain numerical data (and remove columns that contain zero or very few values). These include the following columns: `Var6`, `Var8`, `Var15`, `Var20`, `Var31`, `Var32`, `Var39`, `Var42`, `Var48`, `Var52`, `Var55`, `Var79`, `Var141`, `Var167`, `Var175`, and `Var185`. Figure 9-10 shows the columns that are excluded.

   b. Apply a math operation that adds 1 to each row. The rationale is that this enables you to distinguish between rows that contain actual 0 for the column vs. the substitution value 0 (when you use the Missing Values Scrubber). Figure 9-11 shows the properties for the **Math Operation** module.

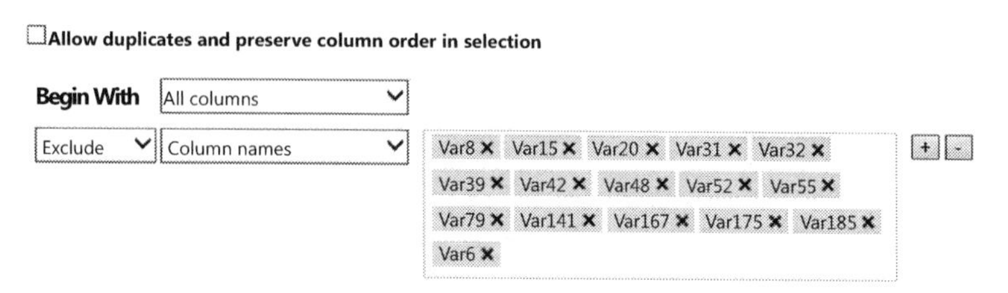

Figure 9-10. *Excluding columns that do not contain useful values*

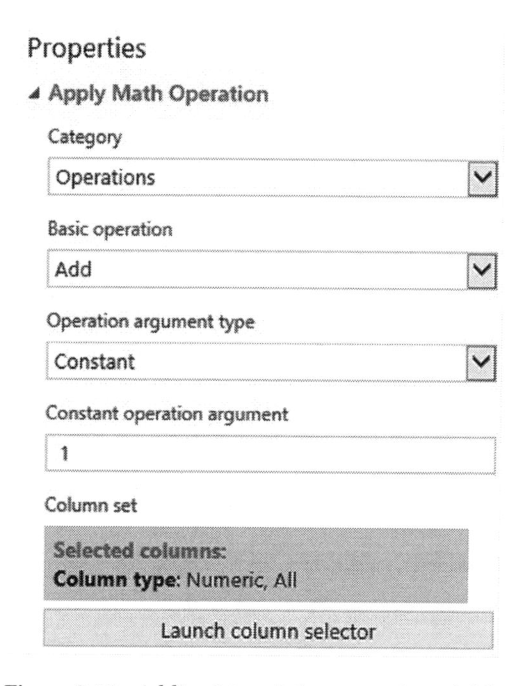

Figure 9-11. *Adding 1 to existing numeric variables*

c. Use the **Missing Value Scrubber** to substitute missing values with 0. Figure 9-12 shows the properties.

Properties
Missing Values Scrubber
For missing values
Custom substitution value
Replace with value
0
Cols with all MV
KeepColumns
MV indicator column
DoNotGenerate

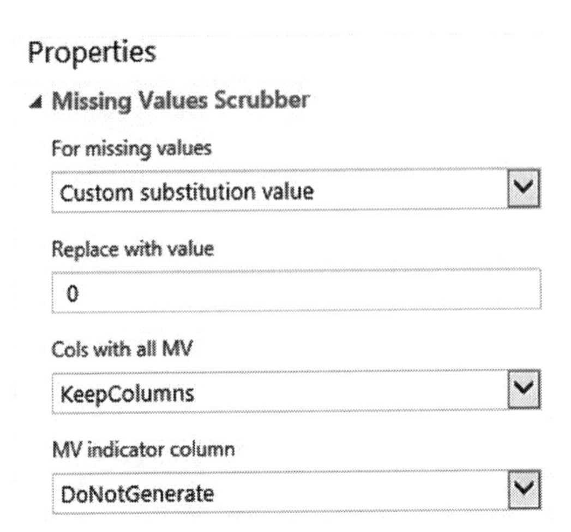

Figure 9-12. *Missing Values Scrubber properties*

d. Use the **Quantize** module to map the input values to a smaller number of bins using a quantization function. In this exercise, you will use the **EqualWidth binning** mode. Figure 9-13 shows the properties used.

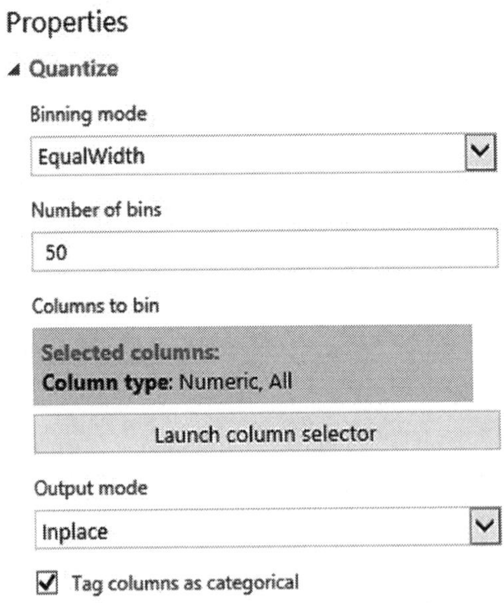

Properties
Quantize
Binning mode
EqualWidth
Number of bins
50
Columns to bin
Selected columns:
Column type: Numeric, All
Launch column selector
Output mode
Inplace
Tag columns as categorical

Figure 9-13. *Quantize properties*

3. For the remaining 40 columns, perform the following steps.

    a. Use the **Missing Values Scrubber** to substitute it with 0. Figure 9-14 shows the properties.

    b. Use the **Metadata Editor** to change the type for all columns to be **categorical**. Figure 9-15 shows the properties.

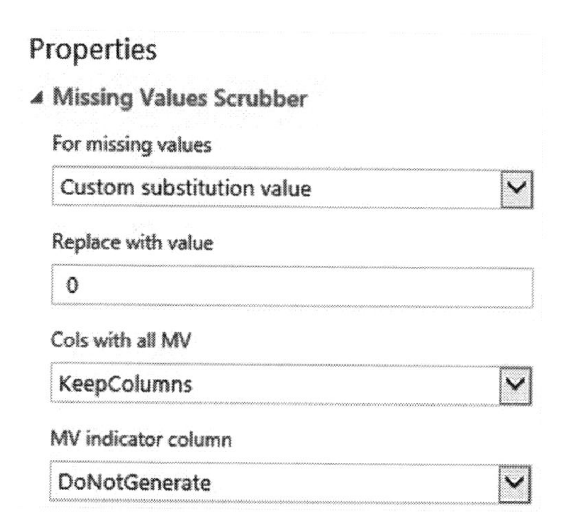

Figure 9-14. *Missing Values Scrubber (for the remaining 40 columns)*

Properties

Metadata Editor

Column

Selected columns:
All columns

Launch column selector

Data type

Unchanged

Categorical

Categorical

Fields

Unchanged

New column names

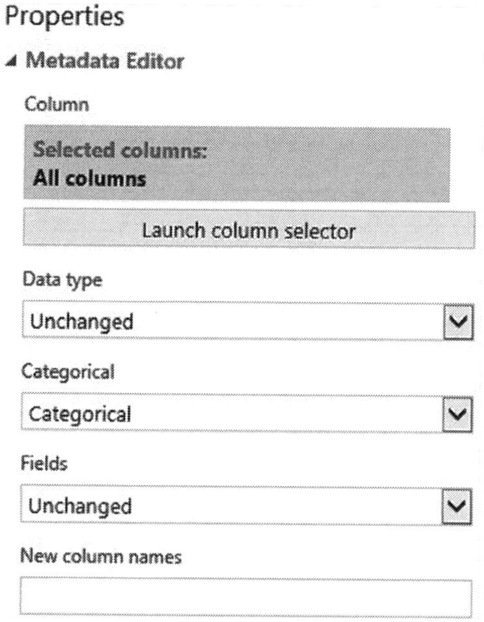

Figure 9-15. *Using the Metadata Editor to mark the columns as containing categorical data*

4. Combine it with the labels from the **ChurnLabel** dataset. Figure 9-16 shows the combined data.

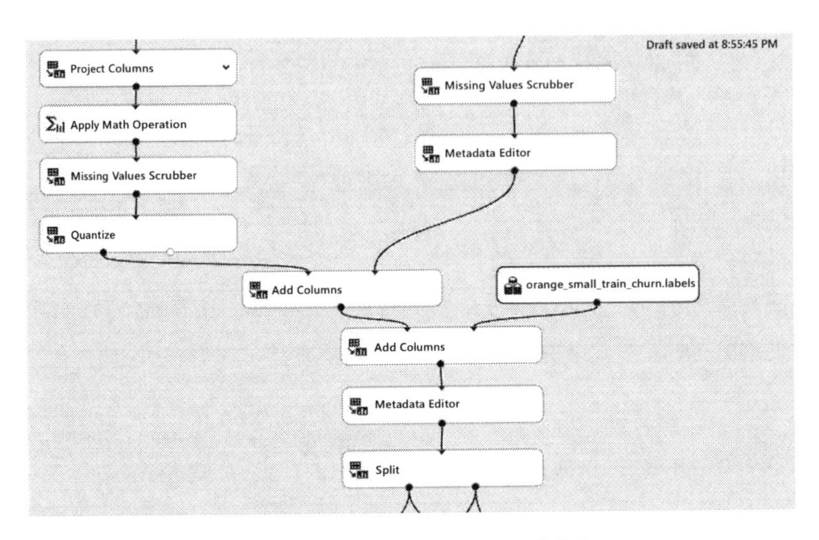

Figure 9-16. *Combining training data and training label*

5. Rename the label column as **ChurnLabel**. Figure 9-17 shows how you can use the Metadata Editor to rename the column.

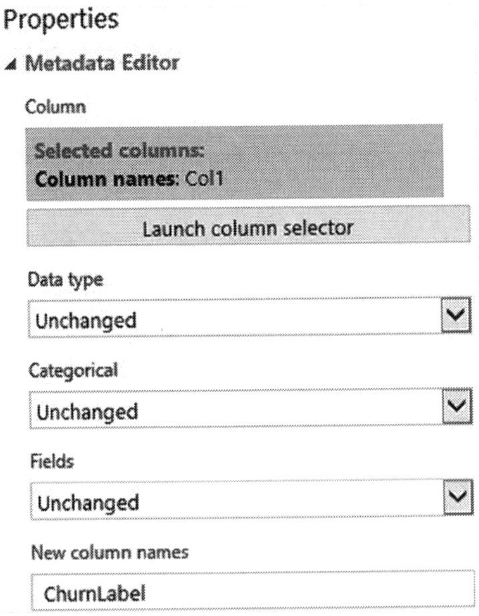

Properties

Metadata Editor

Column

Selected columns:
Column names: Col1

Launch column selector

Data type

Unchanged

Categorical

Unchanged

Fields

Unchanged

New column names

ChurnLabel

Figure 9-17. *Renaming the label column as ChurnLabel*

Classification Model for Predicting Churn

In this section, you will start building the customer churn model using the classification algorithms provided in Azure Machine Learning Studio. For predicting customer churn, you will use two classification algorithms, a two-class boosted decision tree and a two-class decision forest.

A decision tree is a machine learning algorithm for classification or regression. During training, it splits the data using the input variables that give the highest information gain. The process is repeated on each subset of the data until splitting is no longer required. The leaf of the decision tree identifies the label to be predicted (or class). This prediction is provided based on a probability distribution.

The boosted decision tree and decision forest algorithms build an ensemble of decision trees and use them for predictions. The key difference between the two approaches is that, in boosted decision tree algorithms, multiple decision trees are grown in a series such that the output of one tree is provided as input to the next tree. This is a boosting approach to ensemble modeling. In contrast, the decision forest algorithm grows each decision tree independently of each other; each tree in the ensemble uses a sample of data drawn from the original dataset. This is the bagging approach of ensemble modeling. See Chapter 6 for more details on decision trees, decision forests, and boosted decision trees. Figure 9-18 shows how the data is split and used as inputs to train the two classification models.

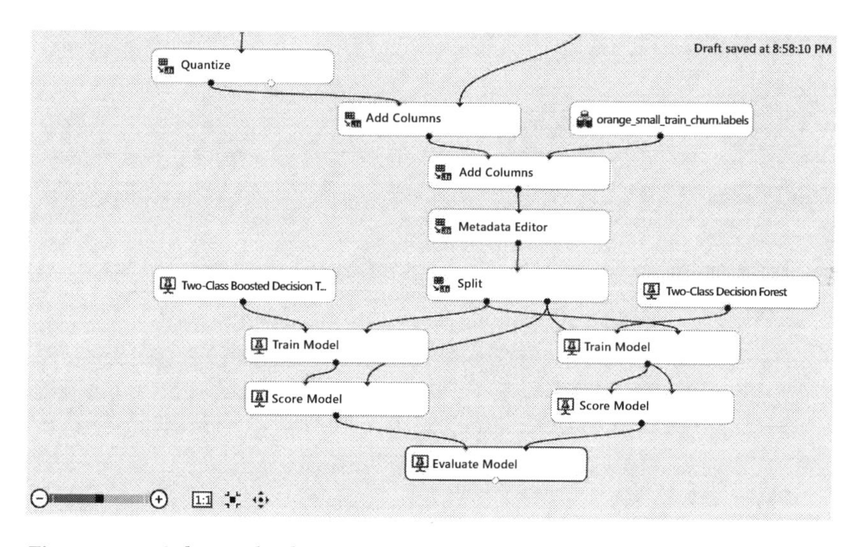

Figure 9-18. *Splitting the data into training and testing, and training the customer churn model*

From Figure 9-18, you can see that the following steps are performed.

1. **Splitting the input data into training and test data**: In this exercise, you split the data by specifying the **Fraction of rows in the first output** dataset, and set it as **0.7**. This assigns 70% of data to the training set and the remaining 30% to the test dataset.

   Figure 9-19 shows the properties for Split.

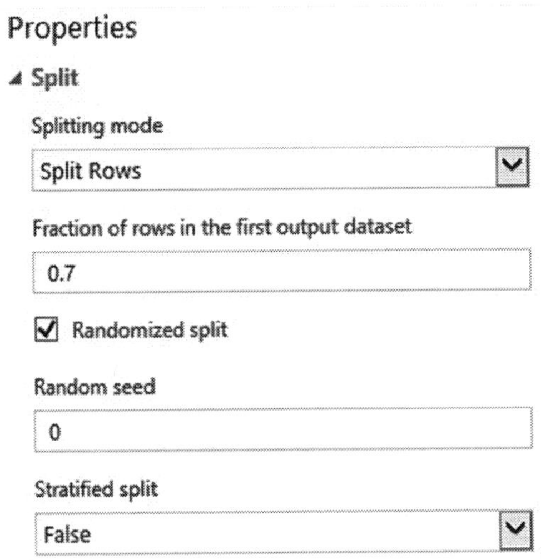

Properties

Split

Splitting mode

Split Rows

Fraction of rows in the first output dataset

0.7

Randomized split

Random seed

0

Stratified split

False

Figure 9-19. *Properties of the Split module*

2. **Training the model using the training data**: In this exercise, you train two classification models, a two-class boosted decision tree and two-class decision forest. Figures 9-20 and 9-21 show the properties for each of the classification algorithms.

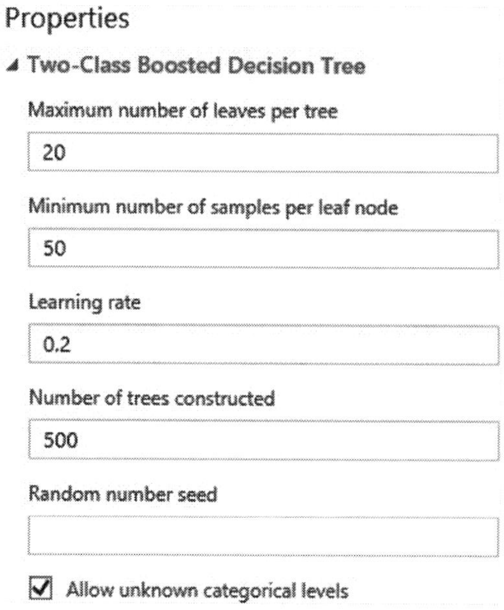

Properties

◢ Two-Class Boosted Decision Tree

Maximum number of leaves per tree

20

Minimum number of samples per leaf node

50

Learning rate

0.2

Number of trees constructed

500

Random number seed

☑ Allow unknown categorical levels

Figure 9-20. *Properties for a two-class boosted decision tree*

Properties

◢ Two-Class Decision Forest

Resampling method

Bagging

Number of decision trees

500

Maximum depth of the decision trees

20

Number of random splits per node

100

Minimum number of samples per leaf node

5

☑ Allow unknown values for categorical features

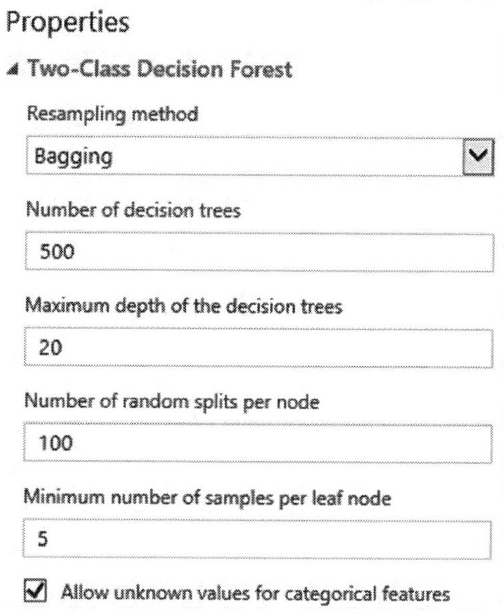

Figure 9-21. *Properties for a two-class decision forest*

3. **Training the model using test data**: To train the model, you need to select the label column. In this exercise, you use the **ChurnLabel** column. Figure 9-22 shows the properties for **Train Model**.

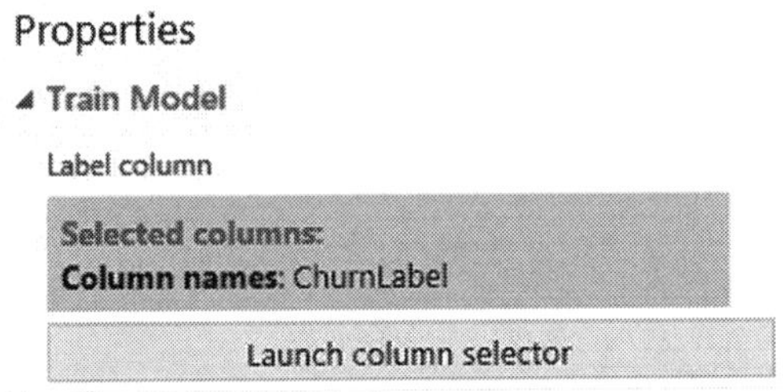

Properties

◢ Train Model

Label column

Selected columns:
Column names: ChurnLabel

Launch column selector

Figure 9-22. Using ChurnLabel as the Label column

Scoring the model: After training the customer churn model, you can use the **Score Model** module to predict the label column for a test dataset. The output of **Score Model** will be used in **Evaluate Model** to understand the performance of the model.

Congratulations, you have successfully built a customer churn model! You learned how to use two of the classification algorithms available in Machine Learning Studio. You also learned how to evaluate the performance of the model. In the next few chapters, you will learn how to deploy the model to production and operationalize it.

Evaluating the Performance of the Customer Churn Models

After you use the Score Model to predict whether a customer will churn, the output of the Score Model module is passed to the Evaluate Model to generate evaluation metrics for each of the model. Figure 9-23 shows the Score Model and Evaluate Model modules.

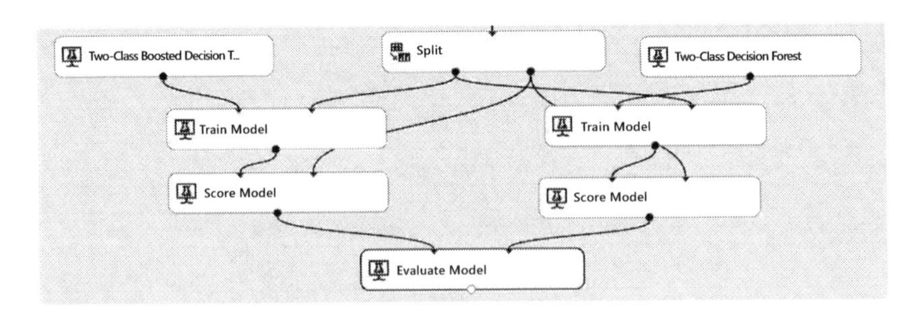

Figure 9-23. Scoring and evaluating the model

After you have evaluated the model, you can right-click the circle at the bottom of the Evaluate Model to see the performance of the two customer churn models. Figure 9-24 shows the Receiver Operating Curve (ROC curve) while Figure 9-25 shows the confusion matrix, accuracy, precision, recall, and F1 scores for the two customer churn models.

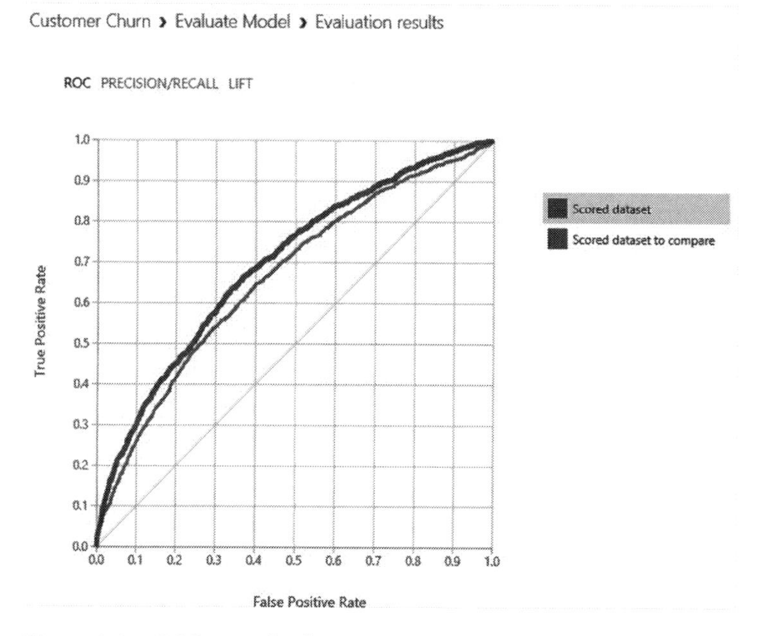

Figure 9-24. *ROC curve for the two customer churn models*

True Positive	False Negative	Accuracy	Precision	Threshold	Cumulative AUC
178	939	0.907	0.283	0.5	0.698
False Positive	**True Negative**	**Recall**	**F1 Score**		
452	13431	0.159	0.204		

Score bin	# Pos	# Neg	Pop.above thresh.	Accuracy	F1	+ve Prec.	+ve Rec.(= TPR)	-ve Prec.	-ve Rec.(= 1 - FPR)	Cumulative AUC
(0.900,1.000]	81	169	0.017	0.920	0.119	0.324	0.073	0.930	0.988	0.001
(0.800,0.900]	32	76	0.024	0.917	0.153	0.316	0.101	0.931	0.982	0.001
(0.700,0.800]	19	66	0.030	0.914	0.169	0.298	0.118	0.932	0.978	0.002
(0.600,0.700]	19	66	0.035	0.910	0.184	0.286	0.135	0.933	0.973	0.002
(0.500,0.600]	26	67	0.041	0.908	0.204	0.285	0.158	0.935	0.968	0.003
(0.400,0.500]	18	104	0.050	0.902	0.210	0.262	0.175	0.935	0.961	0.004
(0.300,0.400]	27	113	0.059	0.896	0.222	0.251	0.199	0.937	0.952	0.006
(0.200,0.300]	27	155	0.071	0.888	0.228	0.234	0.223	0.938	0.941	0.008
(0.100,0.200]	58	348	0.098	0.868	0.237	0.209	0.275	0.940	0.916	0.014
(0.000,0.100]	810	12719	1.000	0.074	0.139	0.074	1.000	1.000	0.000	0.692

Figure 9-25. *Accuracy, precision, recall, and F1 scores for the customer churn models*

The ROC curve shows the performance of the customer churn models. The diagonal line from (0,0) to (1,1) on the chart shows the performance of random guessing. For example, if the yield is randomly selected, the curve will be on the diagonal line. A good predictive model should perform better than random guessing, and the ROC curve should be above the diagonal line. The performance of a customer churn model can be measured by considering the area under the curve (AUC). The higher the area under the curve, the better the model's performance. The ideal model will have an AUC of 1.0, while a random guess will have an AUC of 0.5.

From the visualization, you can see that the customer churn models have a cumulative AUC, accuracy, and precision of 0.698, 0.907, and 0.283, respectively. You can also see that the customer models have a F1 score of 0.204.

■ **Note** See `http://en.wikipedia.org/wiki/F1_score` for a good discussion on the use of the F1 score to measure the accuracy of the machine learning model.

Summary

Using the KDD Cup 2009 Orange telecommunication dataset, you learned step by step how to build customer churn models using Azure Machine Learning. Before building the model, you took time to first understand the data and perform data preprocessing. Next, you learned how to use the two-class boosted decision tree and two-class decision forest algorithms to perform classification, and to build a model for predicting customer churn with the telecommunication dataset. After building the model, you also learned how to measure the performance of the models.

CHAPTER 10

Customer Segmentation Models

In this chapter, you will learn how to build customer segmentation models in Microsoft Azure Machine Learning. Using a practical example, we will present a step-by-step guide to using Microsoft Azure Machine Learning to easily build segmentation models using k-means clustering. After the models have been built, you will learn how to perform validation and deploy it in production.

Customer Segmentation Models in a Nutshell

In order for companies to compete effectively, and build products and services that sell, it is very important to figure out the target customer segments and the characteristics of each segment. Identifying customer segments is critical since it helps companies to better target their marketing campaigns to win the most profitable customers. Data analysts in companies are tasked with sifting through data from both internal and external data sources to identify the magical ingredients that will appeal to specific customer segments (which might not be known a priori).

Customer segmentation empowers companies with the ability to craft marketing strategies and execute marketing campaigns that target and appeal to specific customer segments. This is important because by understanding the customer segments, and what appeals, it leads to greater customer satisfaction.

Note Learn more about market segmentation at `http://en.wikipedia.org/wiki/Market_segmentation`.

Companies must be able to master the deluge of information that is available (such as market research reports, concept/market testing, etc.). While this market research data provides a good balance of qualitative and quantitative information on the market, companies can compete even more effectively if they can tap the huge treasure trove of data that they already have (such as membership/loyalty program databases, billing data from online services/retail outlets, CRM systems, etc.). Somewhere in the treasure trove are insights that companies can turn to their competitive advantage. These insights enable companies to be cognizant of potential customer segments and their characteristics.

For example, in the United States, many people are familiar with the use of the consumer credit score. The consumer credit score helps banks understand the risk profile of customers applying for loans (such as auto and mortgage loans, etc.). This in turn enables banks to tune their interest rates based on the risk segment to which an individual belongs.

Another example is the telecommunication industry. Many telecommunication providers strive to gain insights on how to sell effectively to their customers falling into the two broad groups, corporate business and consumers. In order to figure out effective marketing strategies for consumers, telecommunication providers are often interested in the profile of the person that uses each service. Folks with similar profiles are grouped together and offered discounts or value-added services that appeal to customers with that profile. A telecommunication provider considers many facts about a customer, including income group, call frequency of friends and family members, number of calls/short messages and when they happened, how they pay their monthly bill (online, self-service kiosks, physical stores), delays in payment, etc.

Amongst the different types of unsupervised machine learning techniques, k-means clustering is a common technique used to perform customer segmentation. In this chapter, you will learn how to use Microsoft Azure Machine Learning to perform k-means clustering on the Wikipedia SP 500 dataset (one of the sample datasets available in Azure Machine Learning Studio).

The Wikipedia SP 500 dataset contains the following information: industry category, and text describing each of the 500 Standard & Poor's (SP500) companies. You will build and deploy a k-means clustering model to perform segmentation of the companies. Due to the large number of features that are extracted from the text, you will execute an R script to perform principal component analysis to identify the top 10 features, which will be used to determine the clusters. After k-means clustering has been performed, companies that are similar to each other (based on features extracted from the companies' descriptive text) will be assigned to the same segment.

■ **Note** See Chapter 6 for an overview of the different statistical and machine learning algorithms. Chapter 6 also covers the categories of clustering algorithms, and how k-means clustering works.

Building and Deploying Your First K-Means Clustering Model

To help you get started building your first k-means clustering model, you will use one of the sample experiments provided in Azure Machine Learning Studio. The experiment uses k-means clustering to perform segmentation of companies in the Standard & Poor's (S&P) 500 list of companies.

After the experiment is successfully executed, you can see that the companies (from different industry categories) have been assigned to different clusters (Figure 10-1). In this example, the k-means algorithm finds three segments labeled 0, 1, and 2 on the x-axis. In each of the square boxes, you will see the number of companies from each category that has been assigned to a cluster. In Cluster 2, you can see that there are about 47 companies (from the categories Materials, Health Care, Industrials, Consumer Discretionary, and Information Technology).

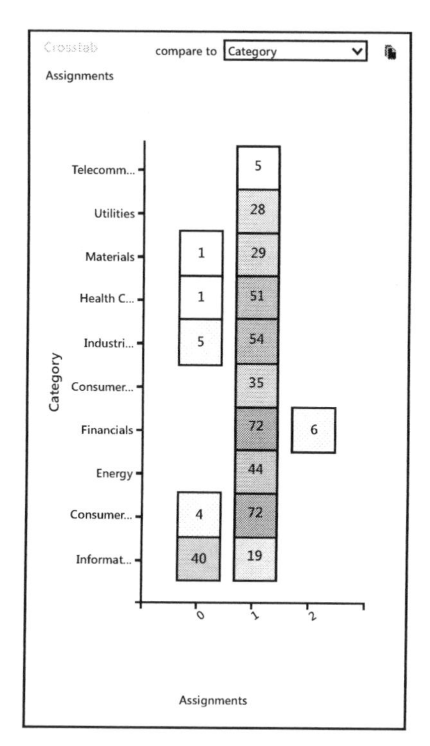

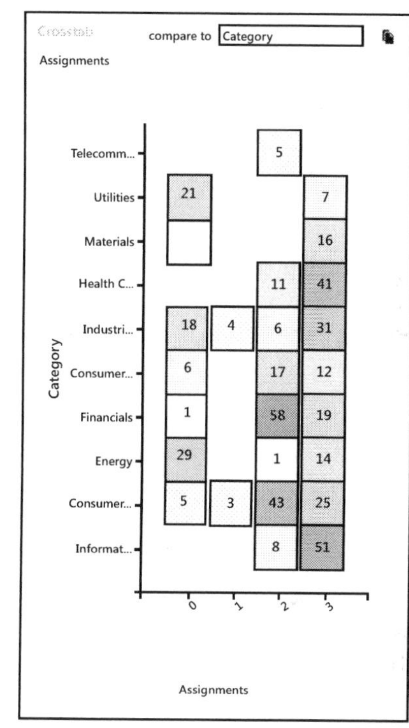

Figure 10-1. *Segmentation of the companies*

Let's get started! For this exercise, let's use the sample experiment called **Clustering: Find similar companies** (shown in Figure 10-2). In Figure 10-3, you can see that the experiment consists of the following steps.

1. Retrieve the data from the Wikipedia SP 500 dataset.
2. Perform feature hashing of the data to vectorize the features.
3. Execute an R script to identify the principal components of the features.
4. Project the columns that will be used in the clustering model.

5. Train the clustering model using two k-means clustering models. For each of the k-means clustering models, different numbers of clusters are specified.

6. Convert the results to a CSV file.

In this section, you will learn how to perform segmentation of the companies.

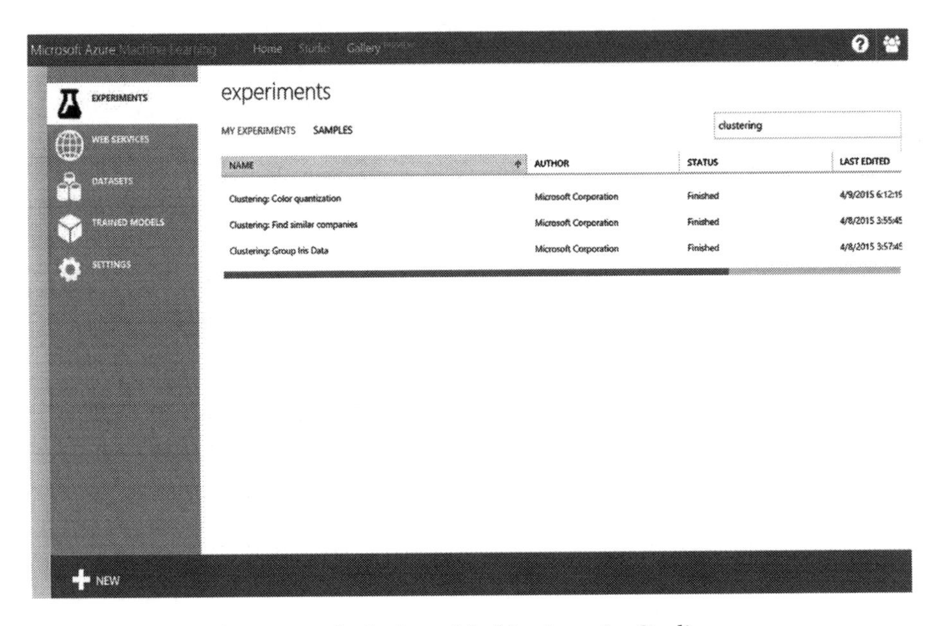

Figure 10-2. Experiment samples in Azure Machine Learning Studio

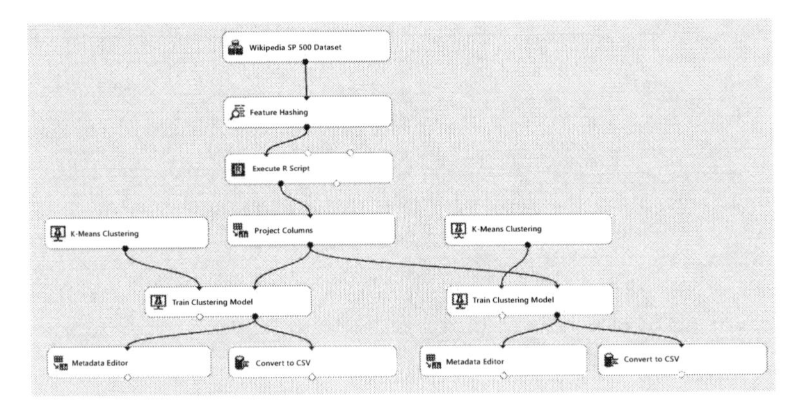

Figure 10-3. *Creating your first k-means clustering model in Azure Machine Learning Studio*

Feature Hashing

In machine learning, the input feature(s) can be free text. A common approach to representing free text is to use a bag of words. Each word is represented as a token. Every time the word appears in the text, a 1 is assigned to the token. And if the word does not appear, a 0 is assigned.

However, the bag-of-words model will not scale because the number of possible words is not known beforehand. Imagine representing a token for every word in the descriptive text of each S&P 500 company. The dimensionality of the inputs can be potentially large. Hence, a common approach is to use a hash function to transform all of the tokens into numerical features, and to restrict the range of possible hash values. To do this, feature hashing (also known as the "hashing trick") is commonly used in machine learning communities to prepare the dataset before it is used as input for machine learning algorithms.

In this example, the Feature Hashing module is used to perform hashing on the descriptive text of the S&P 500 companies. Underneath the hood, the Feature Hashing module uses the Vowpal Wabbit library to perform 32-bit murmurhash hashing. For this exercise, the hashing bit size is set to **12**, and **N-grams**. After the Feature Hashing module has executed, you will see that the descriptive text has been converted to 97 features (shown in Figure 10-4.)

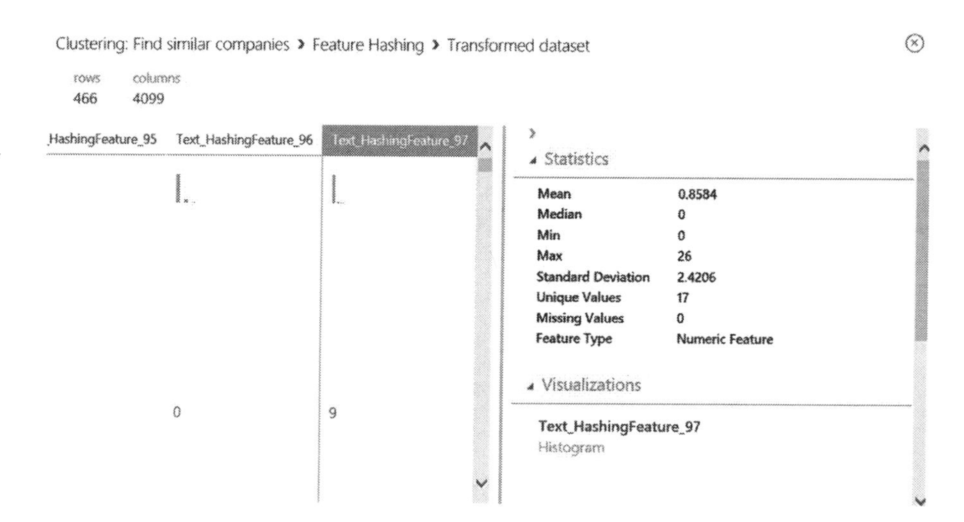

Figure 10-4. *Feature hashing*

■ **Note** The Vowpal Wabbit library is an open source machine learning library. See `https://github.com/JohnLangford/vowpal_wabbit/wiki/Tutorial` for more information.

MurmurHash is a family of non-cryptographic hash functions that provide good distribution, avalanche behavior, and collision resistance. See `https://code.google.com/p/smhasher/` for more information on the MurmurHash family of hash functions.

Identifying the Right Features

When using a k-means cluster for customer segmentation, you need to identify the features that will be used during the clustering process. After feature hashing has been performed, a large number of features will have been computed based on the descriptive text.

Principal Components Analysis (PCA) is a powerful statistical technique that can be used for identifying a smaller number of features (principal components) that captures the key essence of the original features.

In this sample experiment, you will learn how to perform principal component analysis using R. Specifically, you will use the Execute R Script module to execute the following R script (which is provided as part of the sample experiment):

```
# Map 1-based optional input ports to variables
dataset1 <- maml.mapInputPort(1) # class: data.frame

# Sample operation
titles_categories = dataset1[,1:2]
pca = prcomp(dataset1[,4:4099])
top_pca_scores = data.frame(pca$x[,1:10])
data.set = cbind(titles_categories,top_pca_scores)

# You'll see this output in the R Device port.

# It'll have your stdout, stderr and PNG graphics device(s).
plot(pca)

# Select data.frame to be sent to the output Dataset port
maml.mapOutputPort("data.set");
```

From the R script, note that the R function `prcomp` is used. The input to `prcomp` includes column 4 to 4099 of dataset1. The computed results are stored in the variable `pca`. For this sample experiment, you obtain the top 10 principal components, and use these as inputs to the k-means clustering algorithm.

After you have run the experiment, you can click the Execute R Script module, and choose to visualize the result dataset. Figure 10-5 shows the result dataset and the top 10 principal components (PC) that are computed.

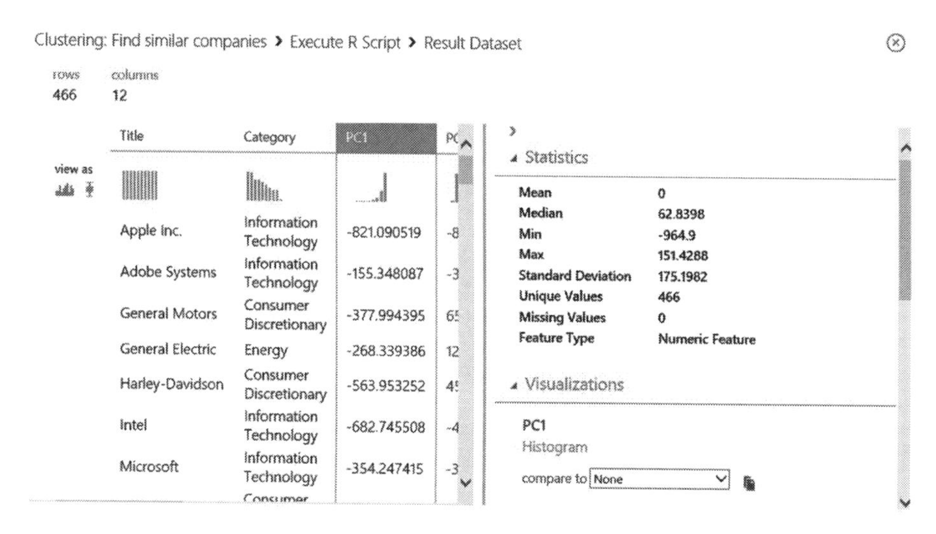

Figure 10-5. *Visualization of the result dataset (top 10 principal components for the dataset)*

■ **Note** Refer to `http://en.wikipedia.org/wiki/Principal_component_analysis` to learn more about PCA.

In R, there are several functions that can be used for performing PCA. These include `pca( )`, `prcomp( )`, and `princomp( )`. `prcomp( )` is commonly used because it is numerically more stable, and it returns an R object with the following information: eigenvectors, square root of the eigenvalues, and scores.

See Chapter 4 for more information on how to use R with Azure Machine Learning.

Properties of K-Means Clustering

In Figure 10-3, you can see that k-means clustering is used twice in the experiment. The main difference between the two K-Means Clustering modules is the number of centroids (value of k). This is also the same as the number of segments or clusters you want from the k-means algorithm. For the left k-means clustering, the sample experiment has specified the value of k to be 3 (the k-means clustering model will identify 3 clusters), whereas the right k-means clustering model specified the value of k to be 5.

You can see the number of centroids specified by clicking each of the k-means clustering rectangles. Figure 10-6 shows the various properties of the k-means clustering model, which include the following:

- **Number of centroids**
- **Metric**: This is the distance metric used to compute the distance between clusters. Azure Machine Learning supports Euclidean and cosine distance metrics.
- **Initialization**: The method used to specify the seed of the initial centroids.
- **Iterations**: The number of iterations used.

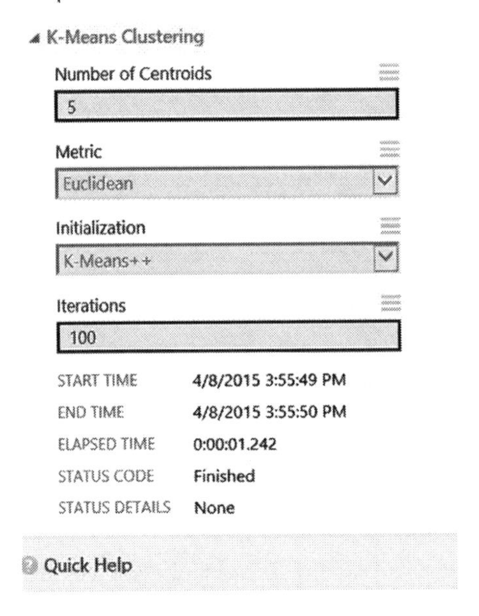

***Figure 10-6.** Properties of the k-means clustering model*

When performing clustering, the user needs to specify the distance measure between any two points in the space. In Azure Machine Learning Studio, this is defined by the Metric property. Two distance measures are supported: Euclidean (also known as L2 norm) and cosine distance. Depending on the characteristics of the input data and the use case, the user should select the relevant Metric property used by k-means clustering.

■ **Note** When performing clustering, it is important to be able to measure the distance (or similarity) between points and vectors. The Euclidean and cosine distances are common distance measures that are used.

Euclidean distance: Given two points, p1 and p2, the Euclidean distance between p1 and p2 is the length of the line segment that connects the two points. The Euclidean distance can also be used to measure the distance between two vectors.

Cosine distance: Given two vectors, v1 and v2, the cosine distance is the cosine of the angle between v1 and v2.

The choice of the distance measure to use is often domain-specific. Euclidean distance is sensitive to the scale/magnitude of the vectors that are compared. It is important to note that even though two vectors can be relatively similar, if the scale of the features are significantly different, the Euclidean distance might show that the two vectors are different. In such cases, cosine distance is often used, since regardless of the scale, the cosine angle between the two vectors would have been small.

After selecting the metric for the distance measure, you need to choose the centroid initialization algorithm. In Azure Machine Learning, this is defined by the `Initialization` property. Five centroid initialization algorithms are supported. Table 10-1 shows the different centroid initialization algorithms.

Table 10-1. *K-Means Cluster, Centroid Initialization Algorithms*

Centroid Initialization Algorithm	Description
Default	Picks first N points as initial centroids
Random	Picks initial centroids randomly
K-Means++	K-means++ centroid initialization
K-Means+ Fast	K-means++ centroid initialization with P:=1 (where the farthest centroid is picked in each iteration of the algorithm)
Evenly	Picks N points evenly as initial centroids

These properties have already been preconfigured for you in the sample experiments.

At this point, you are ready to run the experiment. Click **Run** on the bottom panel of Azure Machine Learning Studio. Once the experiment has successfully executed, two sets of clusters have been produced. The Metadata Editor module is used to change the metadata associated with columns in the dataset to include the assigned cluster. In addition, the Convert to CSV module is used to convert the results to comma-separate values, which allows you to download the result set.

Congratulations! You have successfully run your first company segmentation experiment using the K-Means Clustering module in Azure Machine Learning Studio.

Customer Segmentation of Wholesale Customers

In the earlier section, you learned the key modules used in the sample experiment (K-Means Clustering, Train Clustering Model) to perform customer segmentation. In this section, you will learn step by step how to build the clustering model to perform customer segmentation for a wholesale customer dataset.

■ **Note** The wholesale customers dataset is available on the UCI Machine Learning Repository. The dataset contains eight columns (referred to as attributes or features) and contains information on the customers of a wholesale distributor, operating in different regions.

The columns include annual spending on fresh milk, grocery, frozen, detergents, paper, and delicatessen products. In addition, it also includes information on the customer channel (hotel, cafe, restaurant or retail). Refer to `http://archive.ics.uci.edu/ml/datasets/Wholesale+customers`.

Loading the Data from the UCI Machine Learning Repository

Let's get started by using a Reader module to retrieve the data from the UCI Machine Repository. To do this, drag and drop a **Reader** module (Data Input and Output) from the toolbox. Next, configure the Reader module to read from the HTTP source, and provide the URL for the dataset: `http://archive.ics.uci.edu/ml/machine-learning-databases/00292/Wholesale%20customers%20data.csv`. Figure 10-7 shows the Reader module and its configuration.

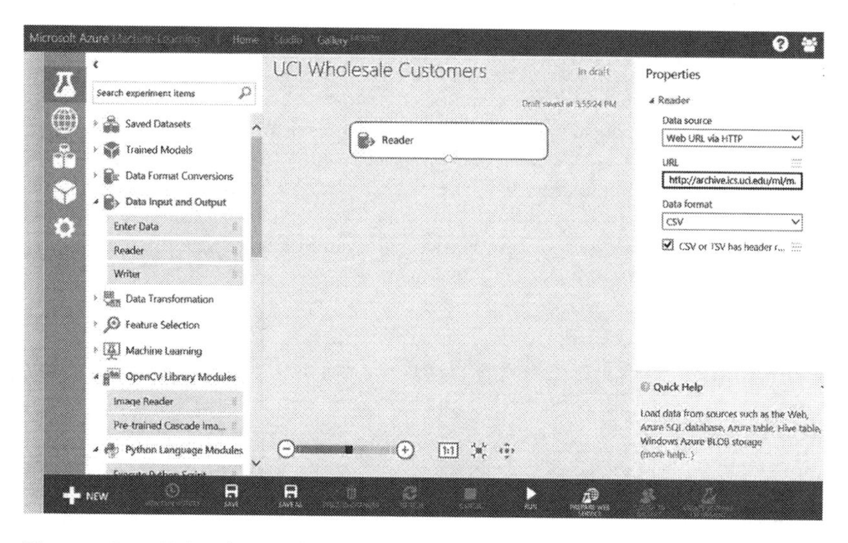

Figure 10-7. *Using the Reader module to read data from the HTTP data source*

Using K-Means Clustering for Wholesale Customer Segmentation

For this experiment, you will use all eight columns from the dataset as inputs for performing clustering. To do this, drag and drop the **K-Means Clustering** and **Train Clustering Model** modules, and connect the modules together based on what is shown in Figure 10-8.

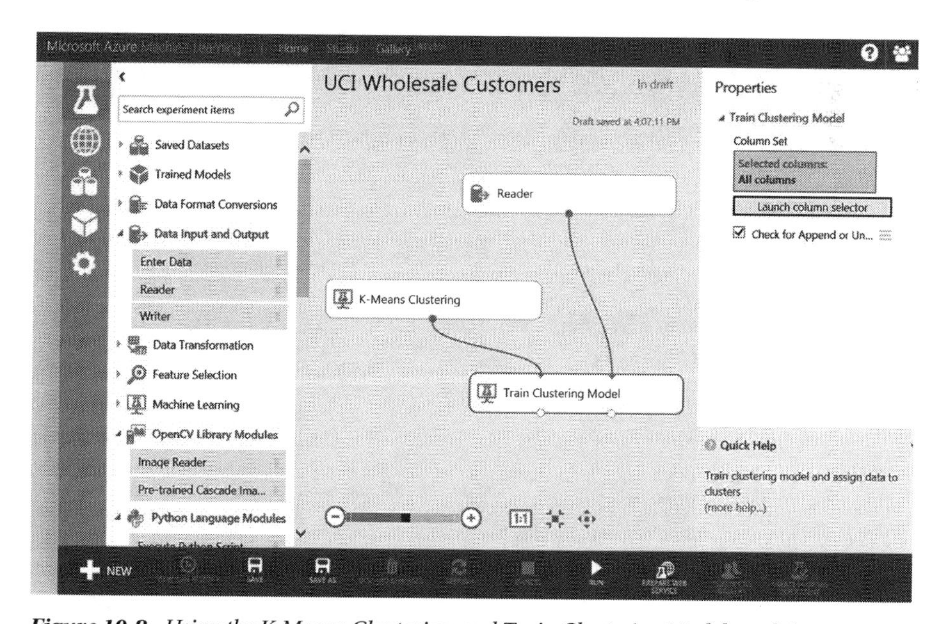

Figure 10-8. *Using the K-Means Clustering and Train Clustering Model modules*

To configure each of the modules, click the module, and specify the values using the **Properties** pane on the right side of the screen. For the K-Means Clustering module, configure it to identify four clusters (**Number of Centroids** = **4**), and use the **Euclidean** distance measure. Use the default **100** number of iterations. For the Train Clustering Mode module, configure it to use **all** of the features when performing clustering.

Finally, you want to visualize the results after the experiment has run successfully. To do this, use the **Metadata Editor** module. Configure the Metadata Editor such that it uses **all** of the features that are produced by the **Train Clustering Model** module. Figure 10-9 shows the final design of the clustering experiment.

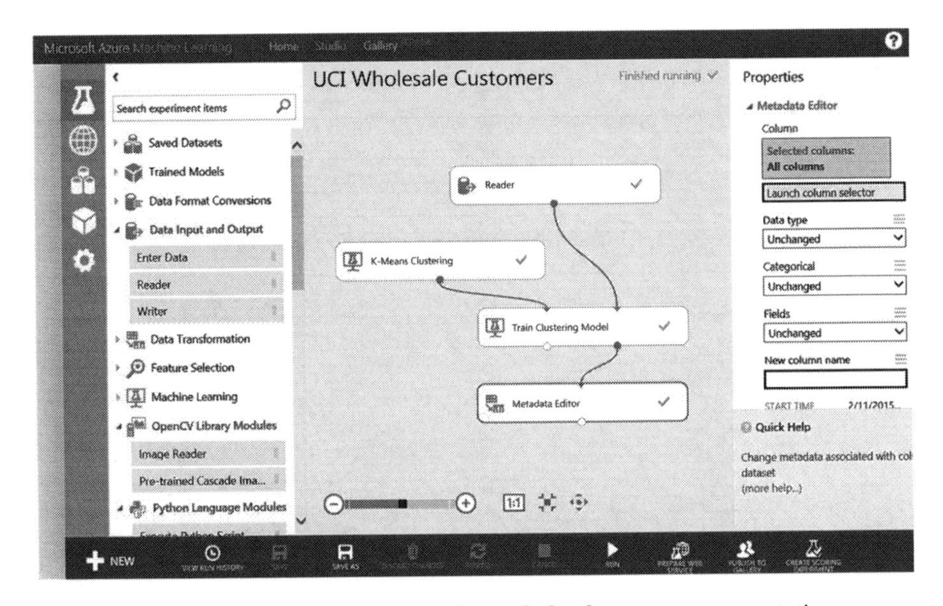

Figure 10-9. *Completed experiment to perform wholesale customer segmentation*

After the experiment has successfully run, you can right-click the **Results** dataset output of the **Metadata Editor** module to see the cluster assignment (shown in Figure 10-10). Scroll to the last column labeled `Assignments`. This shows the cluster assignment for each row in the dataset. Figure 10-10 also shows a chart with the number of records assigned to each cluster. You can see that cluster 2 has the highest frequency with 252 records, while cluster 0 is the smallest with only 26 records.

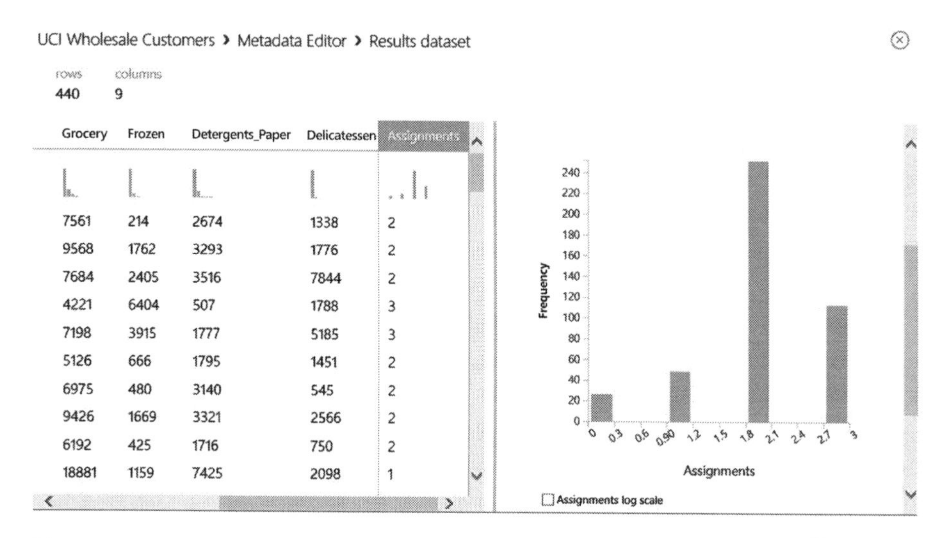

Figure 10-10. *Histogram showing cluster assignments for wholesale customers*

Cluster Assignment for New Data

What happens if you have new customers, and you want to assign them to one of the clusters that you identified? In this section, you will learn how to use the Assign to Cluster module.

In this example, you will first split the input dataset into two sets. The first set will be used for training the clustering model, and the second set will be used for cluster assignments. For practical use cases, the second dataset will be data that you have newly acquired (such as new customers whom you want to assign to a cluster). To split the input dataset, use the **Split** module (in **Data Transformation** > **Sample and Split**), and configure it such that it redirects **90%** of the **input data to its first output**, and the **remaining 10%** to the **second output**. Use the data from the first output of the Split module to train the clustering model.

Figure 10-11 shows the modified clustering experiment design where you added the Assign to Cluster and Split modules. To configure the Assign to Cluster module, click the module, and click the **Launch** column selector to select **all columns. Link** the output of the Assign to Cluster module to the Metadata Editor module. When a new observations is made, it is assigned to the cluster whose centroid has the closest distance.

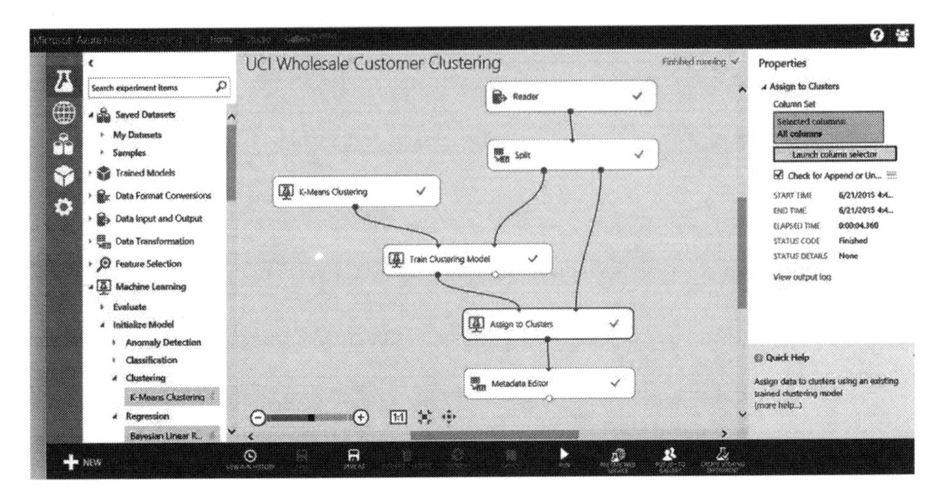

Figure 10-11. *Experiment to perform cluster assignment using the trained clustering model*

You are now ready to run the experiment to perform cluster assignment. After the experiment has run successfully, you can visualize the results by right-clicking the **Results** dataset of the Metadata Editor module, and choosing **Visualize**.

Congratulations! You have successfully built a k-means clustering model, and used the trained model to assign new data to clusters.

Summary

In this chapter, you learned how to create a k-means clustering model using Azure Machine Learning. To jumpstart the learning, you used the sample experiment available in Azure Machine Learning, which performs segmentation of S&P 500 companies (based on each companies' descriptive text). You learned about the key modules that are available in Azure Machine Learning Studio for performing clustering: K-Means Clustering and the Train Clustering Model. You learned how to use feature hashing to vectorize the input (which consists of free text), and PCA to identify the principal components.

In addition, you learned the steps to perform k-means clustering on a wholesale customer dataset, and how to use the trained model to perform cluster assignments for new data.

CHAPTER 11

Building Predictive Maintenance Models

Leading manufacturers are now investing in predictive maintenance, which holds the potential to reduce cost, increase margin and customer satisfaction. Though traditional techniques from statistics and manufacturing have helped, the industry is still plagued by serious quality issues and the high cost of business disruption when components fail. Advances in machine learning offer a unique opportunity to reduce cost and improve customer satisfaction. This chapter will show how to build models for predictive maintenance using Microsoft Azure Machine Learning. Through examples we will demonstrate how you can use Microsoft Azure Machine Learning to build, validate, and deploy a predictive model for predictive maintenance.

Overview

According to Ahmet Duyar, an expert in fault detection and former Visiting Researcher at NASA, a key cause of reduced productivity in the manufacturing industry is low asset effectiveness that results from equipment breakdowns and unnecessary maintenance interventions. In the US alone, the cost of excess maintenance and lost productivity is estimated at $740B, so we clearly need better approaches to maintenance (Duyar, 2011).

Predictive maintenance techniques are designed to predict when maintenance should be performed on a piece of equipment even before it breaks down. By accurately predicting the failure of a component you can reduce unplanned downtime and extend the lifetime of your equipment. Predictive maintenance also offers cost savings since it increases the efficiency of repairs: an engineer can target repair work to the predicted failure and complete the work faster as a result; he/she doesn't need to spend too much time trying to find the cause of the equipment failure. With predictive maintenance, plant operators can be more proactive and fix issues even before their equipment breaks down.

It is worth clarifying the difference between predictive and preventive maintenance since the two are often confused. Preventive maintenance refers to scheduled maintenance that is typically planned in advance. While useful and in many cases necessary, preventive maintenance can be expensive and ineffective at catching issues that develop in between scheduled appointments. In contrast, predictive maintenance aims to predict failures before they happen.

Let's use car servicing as an example to illustrate the difference. When you buy a car, the dealer typically recommends regular services based on time or mileage. For instance, some car manufacturers recommend a full service after 6,000 and 10,000 miles. This is a good example of preventive maintenance. As you approach 6,000 miles, your dealer will send a reminder for you to schedule your full service. In contrast, through predictive maintenance, many car manufacturers would prefer to monitor the performance of your car on an ongoing basis through data relayed by sensors from your car to a database system. With this data, they can detect when your transmission or timing belt are beginning to show signs of impending failure and will proactively call you for maintenance, regardless of your car's mileage.

Predictive maintenance uses non-destructive monitoring during the normal operation of the equipment. Sensors installed on the equipment collect valuable data that can be used to predict and prevent failures.

Current techniques for predictive maintenance include vibration analysis, acoustical analysis, infrared monitoring, oil analysis, and model-based condition modeling. Vibration analysis uses sensors such as accelerometers installed on a motor to determine when it is operating abnormally. According to Ahmet Duyar, vibration analysis is the most widely used approach to condition monitoring, accounting for up 85% of all systems sold. Acoustical analysis uses sonic or ultrasound analysis to detect abnormal friction and stress in rotating machines. While sonic techniques can detect problems in mechanical machines, ultrasound is more flexible and can detect issues in both mechanical and electrical machines. Infrared analysis has the widest range of applications, spanning low- to high-speed equipment as well as mechanical and electrical devices.

Model-based condition monitoring uses mathematical/statistical models to predict failures. First developed by NASA, this technique has a learning phase in which it learns the characteristics of normal operating conditions. Upon completion of the learning phase, the system enters the production phase where it monitors the equipment's condition. It compares the performance of the equipment to the data collected in the learning phase, and will flag an issue if it detects a statistically significant deviation from the normal operation of the machine. This is a form of anomaly detection where the monitoring system flags an issue when the machine deviates significantly from normal operating conditions.

■ **Note** Refer to the following resources for more details on predictive maintenance:

`http://en.wikipedia.org/wiki/Predictive_maintenance` and

Ahmet Duyar, "Simplifying predictive maintenance", Orbit Vol. 31 No.1, pp. 38-45, 2011.

Predictive Maintenance Scenarios

All predictive maintenance problems are not the same. Hence you need to understand the business problem in order to find the most appropriate technique to use. One size does not fit all! Some of the common scenarios include the following:

- Predicting **if** a given component will fail or not before it happens. This allows the manufacturer to proactively repair the component before it fails, which in turn reduces unplanned downtime and cost.
- Diagnosing the most likely causes of failure. With this information the manufacturer can be smarter about the repairs. Engineers can fix the problem much faster if they know the most likely causes. And the repair job will be cheaper as a result.
- Predicting **when** a component will fail. Unlike the first scenario, we need know not just whether a component will fail, but also when the failure is likely to occur. This is important for defining the terms of Service Level Agreements (SLAs) or warranties.
- Predicting the yield failure on a manufacturing plant. This is the subject of this chapter, so let's explore this more in the next section.

The Business Problem

Imagine that you are a data scientist at a semiconductor manufacturer. Your employer wants you to do the following:

- Build a model that predicts yield failure on their manufacturing process, and
- Through your analysis determine the factors that lead to yield failures in their process.

This is a very important business problem for semiconductor manufacturers since their process can be complex, involving involves several stages from raw sand to the final integrated circuits. Given the complexity, there are several factors that can lead to yield failures downstream in the manufacturing process. Identifying the most important factors helps process engineers improve the yield, and reduce error rates and cost of production, leading to increased productivity.

■ **Note** You will need to have an account on Azure Machine Learning. Please refer to Chapter 2 for instructions to set up your new account if you do not have one yet.

The model we will discuss in this chapter is published as the *Predictive Maintenance Model* in the Azure Machine Learning Gallery, which you can access at `http://gallery.azureml.net/`. Feel free to download this experiment to your workspace in Azure Machine Learning.

In Chapter 1, you saw that the data science process typically follows these five steps.

1. Define the business problem
2. Data acquisition and preparation
3. Model development
4. Model deployment
5. Monitoring model performance

Having defined the business problem, you will explore data acquisition and preparation, model development, evaluation, and deployment in the remainder of this chapter.

Data Acquisition and Preparation

Let's explore how to load data from source systems and explore the data in Azure Machine Learning.

The Dataset

For this exercise, you will use the SECOM dataset from the University of California at Irvine's machine learning database. This dataset from the semiconductor manufacturing industry was provided by Michael McCann and Adrian Johnston.

■ **Note** The original SECOM dataset is available at the University of California at Irvine's Machine Learning Repository. This dataset contains 1,567 examples, each with 591 features. Of the 1567 examples, 104 of them represent yield failures.

The features or columns represent sensor readings from 591 points in the manufacturing process. In addition, it also includes a timestamp and the yield result (a simple pass or fail) for each example. In this donated data, the donors did not provide the actual names of each variable. However, they explain that these variables are collected from sensors or process measurement points in the semiconductor manufacturing plant. These variables include both signal and noise. Your task is to determine which of these factors lead to downstream failures in the process. More details are available at `http://archive.ics.uci.edu/ml/machine-learning-databases/secom/secom.names`.

For the full dataset also refer to `http://archive.ics.uci.edu/ml/machine-learning-databases/secom/`.

The data is also available in the *Predictive Maintenance Model* in the Azure Machine Learning Gallery. If you download this experiment to your workspace, you will also get the dataset automatically, so you can skip the data loading section.

Data Loading

Azure Machine Learning enables you to load data from several sources including your local machine, the Web, SQL Database on Azure, Hive tables, or Azure Blob storage.

For this project, we recommend downloading the *Predictive Maintenance Model* from the Azure Machine Learning Gallery to your own workspace since it comes with the required dataset. To learn how to load data into Azure Machine Learning, refer to Chapters 2 and 7.

Data Analysis

Figure 11-1 shows the first part of the experiment that covers data preparation. The data is loaded and preprocessed for missing values using the Clean Missing Data module. Following this, summary statistics are obtained and the Filter-Based Feature Selection module is used to determine the most important variables for prediction.

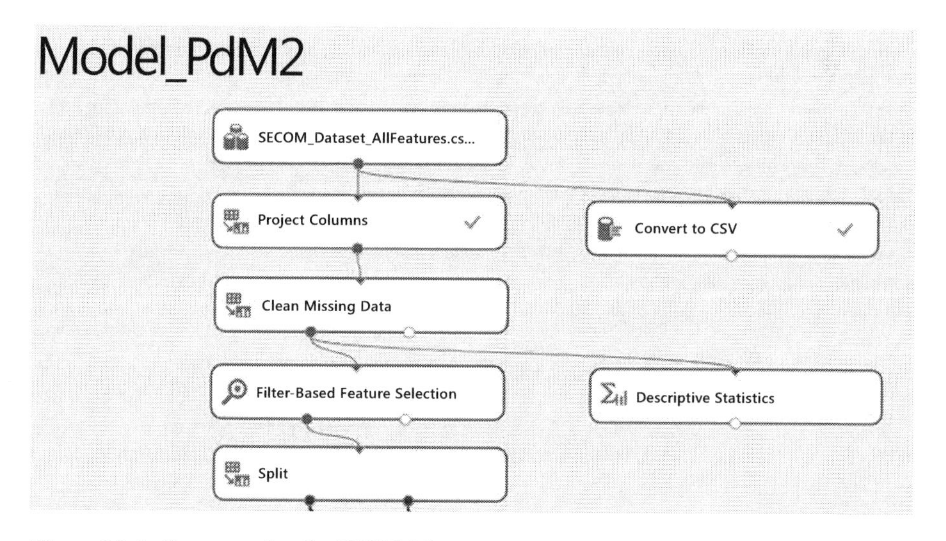

Figure 11-1. *Preprocessing the SECOM dataset*

Figure 11-2 shows a snapshot of the SECOM data as seen in Azure Machine Learning. As you can see, there are 1,567 rows and 592 columns in the dataset. In addition, Azure Machine Learning also shows the number of unique values per feature and the number of missing values. You can see that many of the sensors have missing values. Further, some of the sensors have a constant reading for all rows. For example, sensor #6 has a reading of 100 for all rows. Such features will have to be eliminated because they add no value to the prediction since there is no variation in their readings.

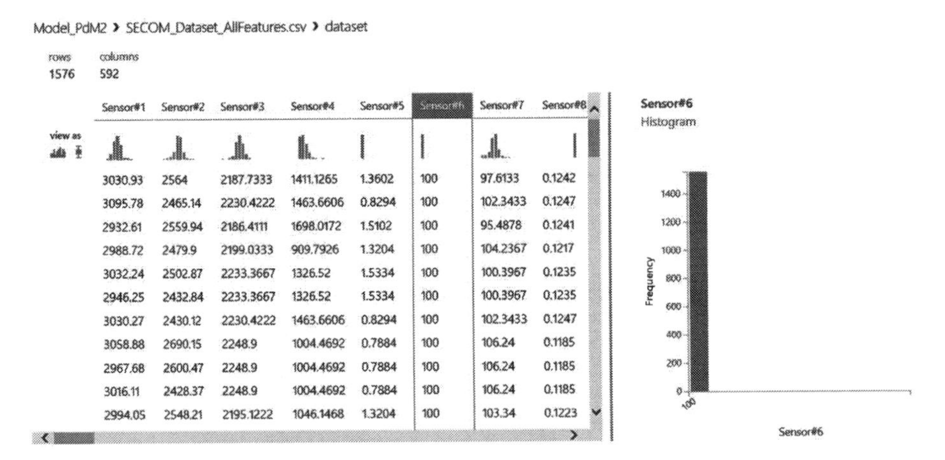

Sensor#1	Sensor#2	Sensor#3	Sensor#4	Sensor#5	Sensor#6	Sensor#7	Sensor#8
3030.93	2564	2187.7333	1411.1265	1.3602	100	97.6133	0.1242
3095.78	2465.14	2230.4222	1463.6606	0.8294	100	102.3433	0.1247
2932.61	2559.94	2186.4111	1698.0172	1.5102	100	95.4878	0.1241
2988.72	2479.9	2199.0333	909.7926	1.3204	100	104.2367	0.1217
3032.24	2502.87	2233.3667	1326.52	1.5334	100	100.3967	0.1235
2946.25	2432.84	2233.3667	1326.52	1.5334	100	100.3967	0.1235
3030.27	2430.12	2230.4222	1463.6606	0.8294	100	102.3433	0.1247
3058.88	2690.15	2248.9	1004.4692	0.7884	100	106.24	0.1185
2967.68	2600.47	2248.9	1004.4692	0.7884	100	106.24	0.1185
3016.11	2428.37	2248.9	1004.4692	0.7884	100	106.24	0.1185
2994.05	2548.21	2195.1222	1046.1468	1.3204	100	103.34	0.1223

Figure 11-2. *Visualizing the SECOM dataset in Azure Machine Learning*

Feature Selection

Feature selection is critical in this case since it provides the answer to the second business problem for this project. Indeed, with over 591 features, you must perform feature selection to identify the subset of features that are useful. Through this process you will also eliminate irrelevant and redundant features. With so many features, you run the risk of the curse of dimensionality. If there is a large number of features, the learning problem can be difficult because the many "extra" features can confuse the machine learning algorithm and cause it to have high variance. It's also important to note that machine learning algorithms are computationally intensive, and reducing the number of features can greatly reduce the time required to train the model, enabling the data scientist to perform experiments in less time. Through careful feature selection, you can find the most influential variables for the prediction. Let's see how to do feature selection in Azure Machine Learning.

To perform feature selection in Azure Machine Learning, drag the module named **Filter-Based Feature Selection** from the list of modules in the left pane. You can find this module by either searching for it in the search box or by opening the **Feature Selection** category. To use this module, you need to connect it to a dataset as the input. Figure 11-3 shows how to use it to perform feature selection on the SECOM dataset. Before running the experiment, use the **Launch column selector** in the right pane to define the target variable for prediction. In this case, choose the column **Yield_Pass_Fail** as the target since this is what you wish to predict. When you are done, set the number of desired features to 50. This instructs the feature selector in Azure Machine Learning to find the top 50 variables.

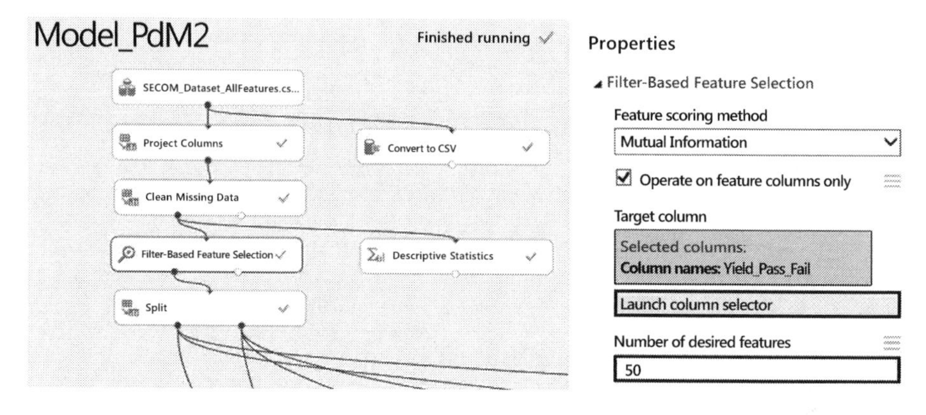

Figure 11-3. *Feature selection in Azure Machine Learning*

You also need to choose the scoring method that will be used for feature selection. Azure Machine Learning offers the following options for scoring:

- Pearson correlation
- Mutual information
- Kendall correlation
- Spearman correlation
- Chi-Squared
- Fischer score
- Count-based

The correlation methods find the set of variables that are highly correlated with the output, but have low correlation among themselves. The correlation is calculated using Pearson, Kendall, or Spearman correlation coefficients, depending on the option you choose.

The Fisher score uses the Fisher criterion from statistics to rank variables. In contrast, the mutual information option is an information theoretic approach that uses mutual information to rank variables. The mutual information option measures the statistical dependence between the probability density of each variable and that of the outcome variable.

Finally, the Chi-Squared option selects the best features using a test for statistical independence; in other words, it tests whether each variable is independent of the outcome variable. It then ranks the variables using a two-way Chi-Squared test.

■ **Note** See `http://en.wikipedia.org/wiki/Feature_selection#Correlation_feature_selection` or

`http://jmlr.org/papers/volume3/guyon03a/guyon03a.pdf` for more information on feature selection strategies.

When you run the experiment, the Filter-Based Feature Selection module produces two outputs: first, the filtered dataset lists the actual data for the most important variables. Second, the module shows a list of the variables by importance with the scores for each selected variable. Figure 11-4 shows the results of the features. In this case, you set the number of features to 50 and you use mutual information for scoring and ranking the variables. Figure 11-4 shows 51 columns since the results set includes the target variable (`Yield_Pass_Fail`) plus the top 50 variables including sensor #60, sensor #248, sensor #520, sensor #104, etc. The last row of the results shows the score for each selected variable. Since the variables are ranked, the scores decrease from left to right.

Note that the selected variables will vary based on the scoring method, so it is worth experimenting with different scoring methods before choosing the final set of variables. The Chi-Squared and mutual information scoring methods produce a similar ranking of variables for the SECOM dataset.

Model_PdM2 › Filter-Based Feature Selection › Features

rows	columns
1	51

	Yield_Pass_Fail	Sensor#60	Sensor#248	Sensor#520	Sensor#1
view as					
	1	0.018432	0.012641	0.012379	0.011491

Figure 11-4. The results of feature selection for the SECOM dataset showing the top variables

Training the Model

Having completed the data preprocessing, the next step is to train the model to predict the yield. Since the response variable is binary, you can treat this as a binary classification problem. So you can use any of the two-class classification algorithms in Azure Machine Learning such as two-class logistic regression, two-class boosted decision trees, two-class decision forest, two-class neural networks, etc.

■ **Note** All predictive maintenance problems are not created equal. Some problems will require different techniques besides classification. For instance, if the goal is to determine ***when*** a part will fail, you will need to use survival analysis. Alternatively, if the goal is to predict energy consumption, you may use a forecasting technique or a regression method that predicts continuous outcomes. Hence you need to understand the business problem in order to find the most appropriate technique to use. One size does not fit all!

Figure 11-5 shows the full experiment to predict the yield from SECOM data. The top half of the experiment, up to the Split module, implements the data preprocessing phase. The Split module splits the data into two samples, a training sample comprising 70% of the initial dataset, and a test sample with the remaining 30%.

In this experiment, you compare three classification algorithms to find the best prediction of yield. The left branch after the Split module uses the two-class boosted decision tree, the middle branch uses the two-class logistic regression algorithm that is widely used in statistics, and the right branch uses the two-class Bayes Point Machine. Each of these algorithms is trained with the same training sample and tested with the same test sample. You use the module named Train Model to train each algorithm and the Score Model for testing. The module named Evaluate Model is used to evaluate the accuracy of these two classification models.

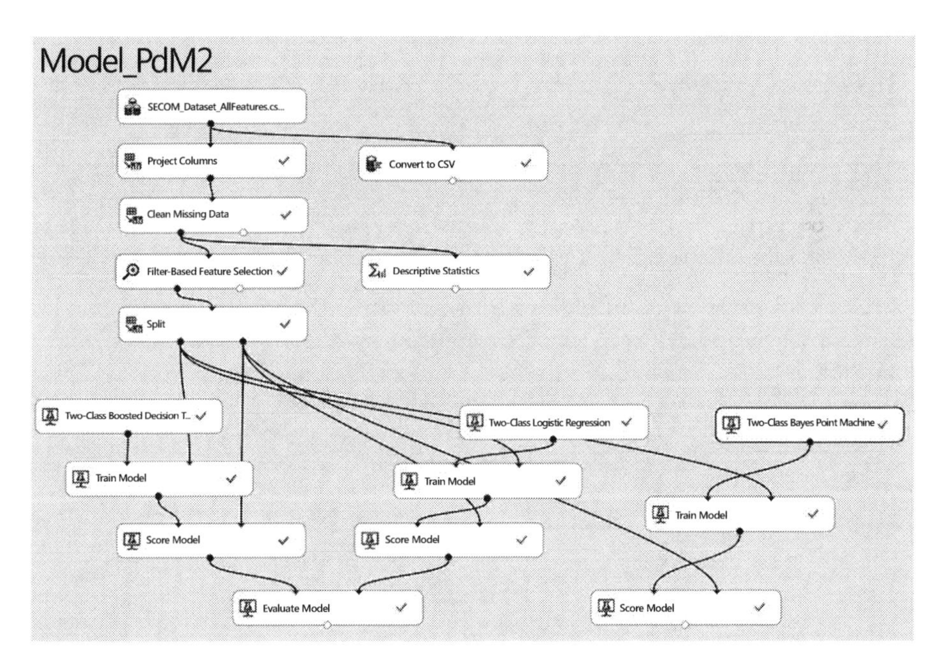

Figure 11-5. An experiment for predicting the yield from SECOM data

Model Testing and Validation

Once the model is trained, the next step is to test it with a hold-out sample to avoid over-fitting and evaluate model generalization. We have shown how we performed this using a 30% sample for testing the trained model. Another strategy to avoid over-fitting and evaluate model generalization is cross-validation, which was discussed in Chapter 7. By default, Azure Machine Learning uses 10-fold cross-validation. With this approach, you cycle through 10 hold-out samples instead of one for validation. To perform cross-validation for this problem you can simply replace any pair of the Train Model and Score Model modules with the module named Cross Validate Model. Figure 11-6 shows how to perform cross-validation with the Two-Class Logistic Regression portion of this experiment. You will also need to use the whole dataset from the Filter-Based Feature Selection as your input dataset. For cross-validation there is no need to split the data into training and test sets with the Split module since the module named Cross Validate Model will do the required data splits automatically.

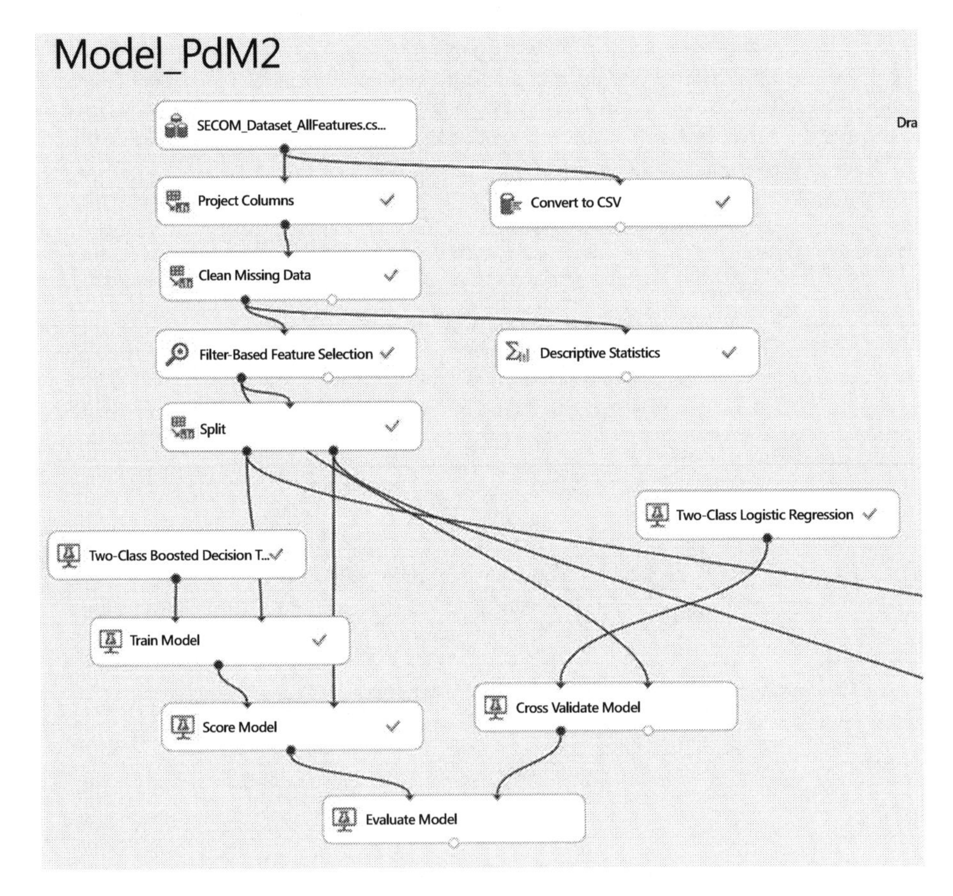

Figure 11-6. A modified experiment with cross-validation

Model Performance

The Evaluate Model module is used to measure the performance of a trained model. This module accepts two datasets as inputs. The first is a scored dataset from a tested model. The second is an optional dataset for comparison. You can use this module to compare the performance of the same model on two different datasets, or to compare the performance of two different models on the same dataset so that you can know which model performs better on the task. After running the experiment, you can check your model's performance by clicking **the small circle** at the bottom of the module named **Evaluate Model**. This module provides the following methods to measure the performance of a classification model such as the propensity model:

- The **receiver operating characteristic**, or ROC curve, which plots the rate of true positives to false positives
- The **lift curve** (also known as the gains curve), which plots the number of true positives vs. the positive rate
- The **Precision vs. recall** chart that shows the model's precision at different recall rates
- The **confusion matrix** that shows type I and II errors

Figure 11-7 shows the ROC curve for the model you built earlier. The ROC curve visually shows the performance of a predictive binary classification model. The diagonal line from (0,0) to (1,1) on the chart shows the performance of random guessing, so if you randomly selected the yield, your predictions would be on this diagonal line. A good predictive model should do much better than random guessing. Hence, on the ROC curve, a good model should lie above the diagonal line. The ideal or perfect model that is 100% accurate will have a vertical line from (0,0) to (0,1), followed by a horizontal line from (0,1) to (1,1).

One way to measure the performance from the ROC curve is to measure the area under the curve (AUC). The higher the area under the curve, the better the model's performance. The ideal model will have an AUC of 1.0, while random guessing will have an AUC of 0.5. The logistic regression model you built has an AUC of 0.758 while the boosted decision tree model has a slightly higher AUC of 0.776! In this experiment, the two-class boosted decision tree had the following parameters:

- Maximum number of leaves per tree = 101 – 900
- Minimum number of samples per leaf = 101 – 900
- Learning rate = 0.1 – 0.9
- Number of trees constructed = 1001 – 9000
- Random number seed = 1

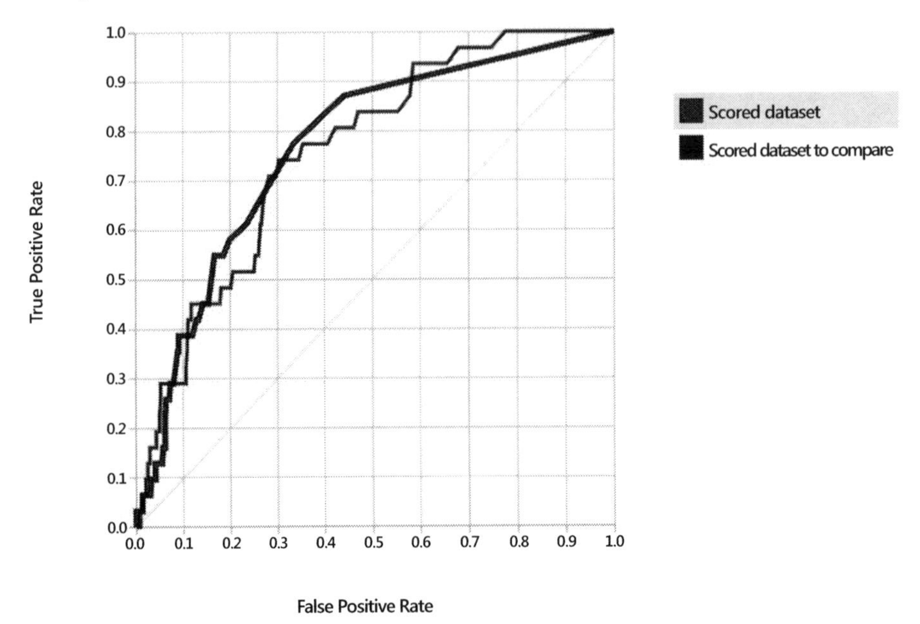

Figure 11-7. *Two ROC curves comparing the performance of logistic regression and boosted decision tree models for yield prediction*

In addition, Azure Machine Learning also provides the confusion matrix as well as precision and recall rates for both models. By clicking each of the two curves, the tool shows the confusion matrix, precision, and recall rates for the selected model. If you click the bottom curve or the legend named "scored dataset to compare," the tool shows the performance data for the logistic regression model.

Figure 11-8 shows the confusion matrix for the boosted decision tree model. The confusion matrix has the following four cells:

- **True positives**: These are cases where yield failed and the model correctly predicts failure.
- **True negatives**: In the historical dataset, the yield passed, and the model correctly predicts a pass.
- **False positives**: In this case, the model incorrectly predicts that the yield would fail when in fact it passed. This is commonly referred to as a Type I error. The boosted decision tree model you built had only three false positives.
- **False negatives**: Here the model incorrectly predicts that the yield would pass when in reality it failed. This is also known as a Type II error. The boosted decision tree model had 30 false negatives.

True Positive	False Negative	Accuracy	Precision	Threshold	AUC
1	30	0.930	0.250	0.5	0.776
False Positive	**True Negative**	**Recall**	**F1 Score**		
3	439	0.032	0.057		

Figure 11-8. *Confusion matrix and more performance metrics*

In addition, Figure 11-8 shows the accuracy, precision, and recall of the model. Here are the formulas for these metrics:

Precision is the rate of true positives in the results.

$$Precision = \frac{tp}{tp + fp} = \frac{1}{1+3} = 0.25$$

Recall is the percentage of yield failures that the model identifies and is measured as

$$Recall = \frac{tp}{tp + fn} = \frac{1}{1+30} = 0.032$$

Finally, the accuracy measures how well the model correctly identifies yield failures and passes, shown as

$$Accuracy = \frac{tp + tn}{tp + tn + fp + fn} = \frac{1+439}{1+439+3+30} = 0.93$$

where tp = true positive, tn = true negative, fp = false positive, and fn = false negative. The F1 score is a weighted average of precision and recall. In this case, it is quite low since the recall is very low.

Techniques for Improving the Model

So given all these metrics, how good is our model? Its accuracy is very high at 93%. Is this good or bad? As you saw in Chapter 7, the accuracy of a model can be misleading especially where there is class imbalance. Please refer to the section entitled "Prioritizing Evaluation Metrics" in Chapter 7 for more details on this.

This specific model is not great. Despite its high accuracy, the model has very low precision, recall, and F1 score. So why is the accuracy is so high? The reason is class imbalance. The SECOM dataset has only 104 cases of yield failure out of 1,567 examples. This is a 6.6% failure rate. With such an imbalanced class even a naïve model can show high accuracy by simply predicting "yield pass" every time. Such a model would have a 93.4% accuracy without correctly identifying a single case of yield failure. With this class imbalance the model learns to predict which cases will pass but fails to learn about yield failures which is the goal of the project.

To address this problem you can use two techniques that will help the classification algorithms to learn the minority class (yield failures): downsampling and upsampling. With downsampling we balance the two classes by reducing the number of cases with yield passes. By reducing the number of passed cases we increase the proportion of yield failures in our sample. This will enable the Machine Learning algorithms to learn to better predict yield failures. With upsampling we use resampling techniques to increase the number of yield failure cases in the data. This also has the effect of increasing the proportion of yield failure cases in our sample, which helps the algorithms to better learn to predict yield failures.

Now let's explore how you can do upsampling and downsampling in Azure Machine Learning.

■ **Note** For more information on upsampling and downsampling refer to the following papers:

Chen, Chao, Liaw, Andy, and Breiman, Leo. "Using Random Forest to learn imbalanced data". University of California at Berkeley, Department of Statistics. Report Number 666, 2004.

Li, Junyi and Nenkova, A. "Addressing Class Imbalance for Improved Recognition of Implicit Discourse Relations". Proceedings of the SIGDIAL 2014 Conference, pp. 142–150, Philadelphia, U.S.A., 18–20 June, 2014.

Chawla, N. V., Bowyer, K. W., Hall, L. O., and Kegelmeyer, W. P. (2002). Smote: Synthetic minority over-sampling technique. Journal of Artificial Intelligence Research, 16:321–357.

The SMOTE algorithm in R is in the DMwR package. To use this algorithm you need to install the DMwR package onto your local machine. Detailed instructions for installing any R package into Azure Machine Learning is available on MSDN at `http://blogs.msdn.com/b/benjguin/archive/2014/09/24/how-to-upload-an-r-package-to-azure-machine-learning.aspx`. We highly recommend this resource if you plan to use the SMOTE implementation from R.

Upsampling and Downsampling

The upsampling technique addresses class imbalance by increasing the number of examples from the minority class. This creates a more balanced dataset with more representation from the minority class. With a more balanced dataset the learning algorithms have a much better chance of learning how to correctly predict the minority class. You can do upsampling easily in Azure Machine Learning using the SMOTE module. This module implements the Synthetic Minority Oversampling TEchnique (SMOTE) developed by Chawla et al. in 2002. This algorithm resolves class imbalance by artificially generating synthetic examples from the minority class using a nearest neighbors approach. The SMOTE algorithm has been implemented in other Machine Learning platforms such as R and WEKA.

In this example, you will use the SMOTE algorithm in R since it does both upsampling and downsampling. You can use this R implementation of the SMOTE algorithm in Azure Machine Learning with the following steps.

First, copy the Predictive Maintenance Model from the Gallery (at `http://gallery.azureml.net/`) to your workspace.

1. The module named **Missing Value Scrubber** is being deprecated, so replace it with the new module named **Clean Missing Data**.
2. Find the module named **Normalize Data** and connect its input to the output of the **Clean Missing Data** module.
3. Connect the output of the **Normalize Data** module to the input of the **Filter-Based Feature Selection** module
4. Drag the new ZIP file containing the DMwR package to the canvas. Also, drag the **Execute R Script** module to the canvas and insert it after the **Split** module.
5. Connect the DMwR package to the third input of the **Execute R Script** module. Also, connect the first output of the **Split** module to the first input of the **Execute R Script** module.
6. Now connect the first output of the **Execute R Script** module to the second input of the Train Model modules for the Boosted Decision Tree, Logistic Regression, and Bayes Point Machine modules. The result is shown in Figure 11-9.
7. Edit the parameters of the **Execute R Script** module as shown in Figure 11-10. The most critical part is line 24 that calls the SMOTE algorithm with the call

   ```
   data.set <- SMOTE(Yield_Pass_Fail ~ ., dataset1, k=3,
   perc.over=300, perc.under=100)
   ```

   The first parameter is the formula for the prediction problem. In this case the notation `Yield_Pass_Fail ~ .` denotes that you are trying to predict `Yield_Pass_Fail` using all input variables.

   `Dataset1` is the training sample passed to the Execute R Script module.

   `k=3` instructs SMOTE to use the three nearest neighbors to generate a new synthetic example from the minority class.

   `perc.over=300` tells SMOTE to create three new failure cases for every example of yield failure in the dataset.

   `perc.under=100` instructs the SMOTE algorithm to select exactly one case from the majority class for every failure case generated. In this case, it will select three passed examples for every three new failure cases it generates.
8. Now run the experiment by clicking the **Run** button at the bottom of Azure Machine Learning Studio.

9. Inspect the results of the experiment as follows.

   a. Right-click the dot at the bottom of the **Evaluate Model** module

   b. Choose **Visualize** from the menu. The results are shown in Figures 11-11 and 11-12.

Figure 11-11 shows the ROC curve while Figure 11-12 shows the confusion matrix with the accuracy, recall, precision, and F1 score. It is evident that the performance of the Boosted Decision Tree and Logistic Regression models improve after oversampling and undersampling with the SMOTE algorithm. Logistic Regression performs better than Boosted Decision Tree after oversampling and undersampling. Although accuracy and precision decrease, recall rate increases from a mere 3.2% to 78.9% while the F1 score also increases from 5.7% to 23.4. Area Under the Curve (AUC) decreases slightly from 77.6% to 75.1%.

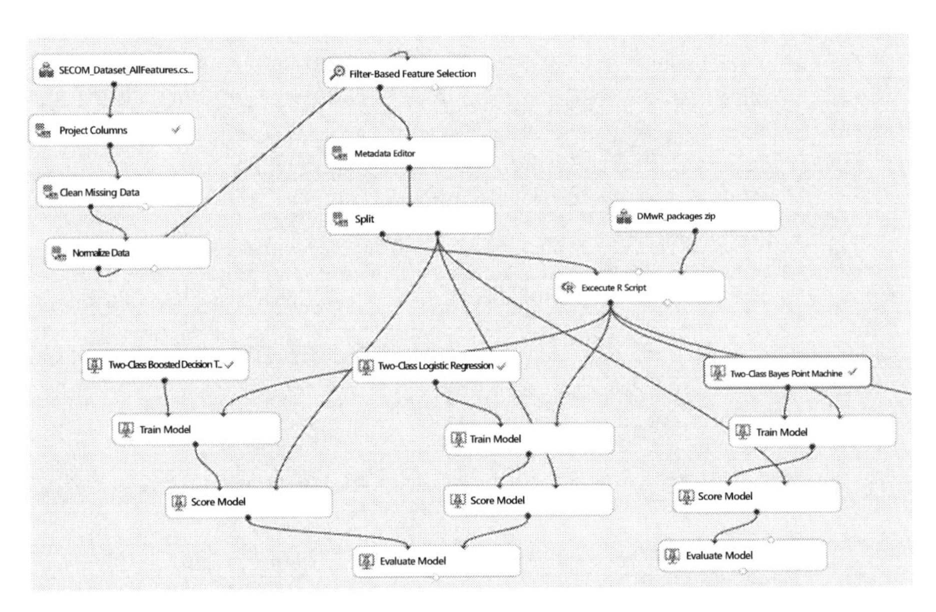

Figure 11-9. *The modified Predictive Maintenance Model with the added Execute R Script module and the DMwR package*

◢ Execute R Script

R Script

```
1  # Map 1-based optional input ports to variables
2  dataset1 <- maml.mapInputPort(1) # class: data.frame
3  #dataset2 <- maml.mapInputPort(2) # class: data.frame
4
5  # Contents of optional Zip port are in ./src/
6  # source("src/yourfile.R");
7  # load("src/yourData.rdata");
8
9  install.packages("src/gtools_3.4.1.zip", lib = ".", repos = NULL, verbose = TRUE)
10 install.packages("src/gdata_2.13.3.zip", lib = ".", repos = NULL, verbose = TRUE)
11 install.packages("src/TTR_0.22-0.zip", lib = ".", repos = NULL, verbose = TRUE)
12 install.packages("src/gplots_2.16.0.zip", lib = ".", repos = NULL, verbose = TRUE)
13 install.packages("src/xts_0.9-7.zip", lib = ".", repos = NULL, verbose = TRUE)
14 install.packages("src/quantmod_0.4-4.zip", lib = ".", repos = NULL, verbose = TRUE)
15 install.packages("src/abind_1.4-3.zip", lib = ".", repos = NULL, verbose = TRUE)
16 install.packages("src/ROCR_1.0-7.zip", lib = ".", repos = NULL, verbose = TRUE)
17 install.packages("src/DMwR_0.4.1.zip", lib = ".", repos = NULL, verbose = TRUE)
18
19 library(DMwR, lib.loc=".", verbose=TRUE)
20
21 dataset1$Yield_Pass_Fail <- as.factor(dataset1$Yield_Pass_Fail)
22
23
24 data.set <- SMOTE(Yield_Pass_Fail ~ ., dataset1, k=3, perc.over = 300, perc.under=100)
25 # Sample operation
26 # data.set = rbind(dataset1, dataset2);
27
```

Figure 11-10. *Invoking the SMOTE algorithm in the DMwR package from the Execute R Script module in Azure Machine Learning*

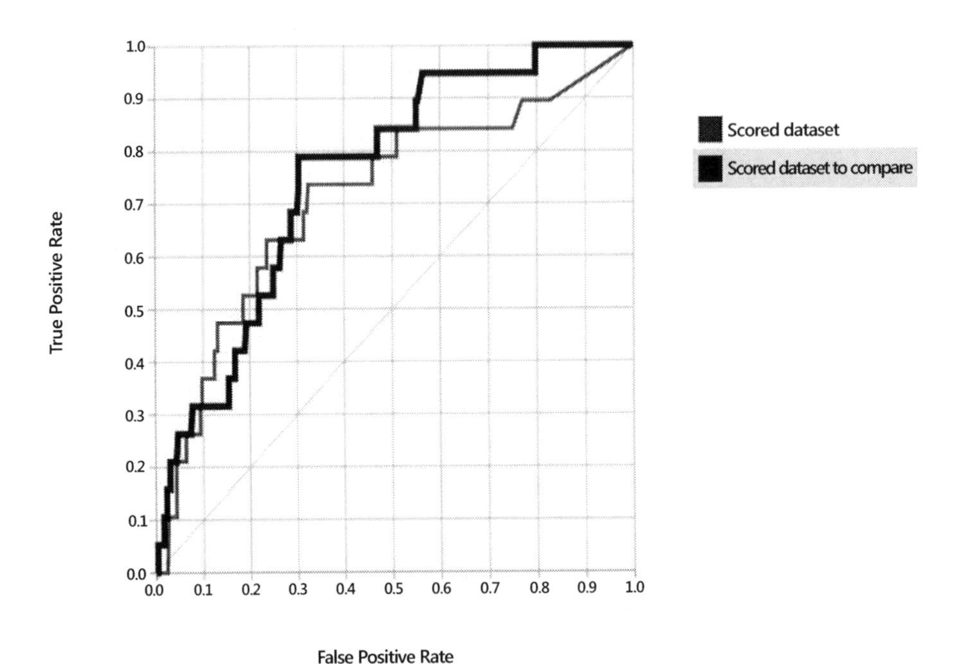

Figure 11-11. *The ROC curves of the Boosted Decision Tree and Logistic Regression models after upsampling and downsampling with the SMOTE algorithm in R*

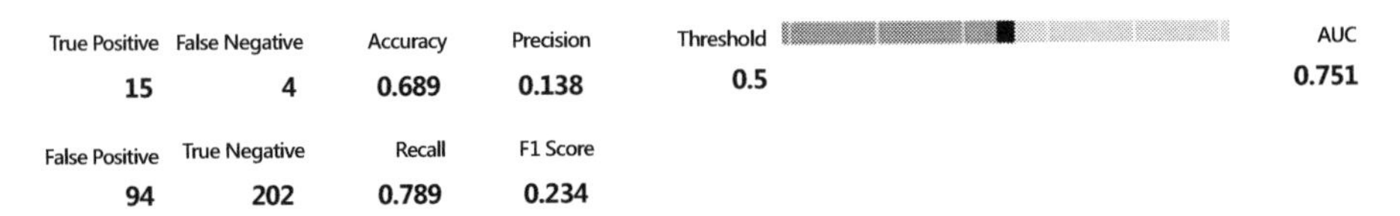

Figure 11-12. *The confusion matrix, precision, recall, and F1 score of the Boosted Decision Tree model after upsampling and downsampling with the SMOTE algorithm in R*

Model Deployment

When you build and test a predictive model that meets your needs, you can use Azure Machine Learning to deploy it into production for business use. A key differentiator of Azure Machine Learning is the ease of deployment in production. Once a model is complete, it can be deployed very easily into production as a web service. Once deployed, the model can be invoked as a web service from multiple devices including servers, laptops, tablets, or even smartphones.

The following two steps are required to deploy a model into production.

1. Create a predictive experiment, and
2. Publish your experiment as a web service.

Let's review these steps in detail and see how they apply to your finished model built in the previous sections.

Creating a Predictive Experiment

To create a predictive experiment, follow these steps in Azure Machine Learning Studio.

1. Run your experiment with the **Run** button at the bottom of Azure Machine Learning Studio.
2. Select the **Train Model** module in the left branch containing the Boosted Decision Tree. This tells the tools that you plan to deploy the Boosted Decision Tree model in production. This step is only necessary if you have several training modules in your experiment.
3. Next, click **Set Up Web Service | Predictive Web Service (Recommended)** at the bottom of Azure Machine Learning Studio. Azure Machine Learning will automatically create a predictive experiment. In the process, it deletes all the modules that are not needed. For example, all the other **Train Model** modules, the **Split**, **Project**, and other modules are removed. The tool replaces the **Two-Class Boosted Decision Tree** module and its **Train Model** module with the newly trained model. It also adds a new web input and output for your experiment.

Your predictive experiment should appear as shown Figure 11-13. You are now ready to deploy your new predictive experiment in production. To do this, let's move to the next step.

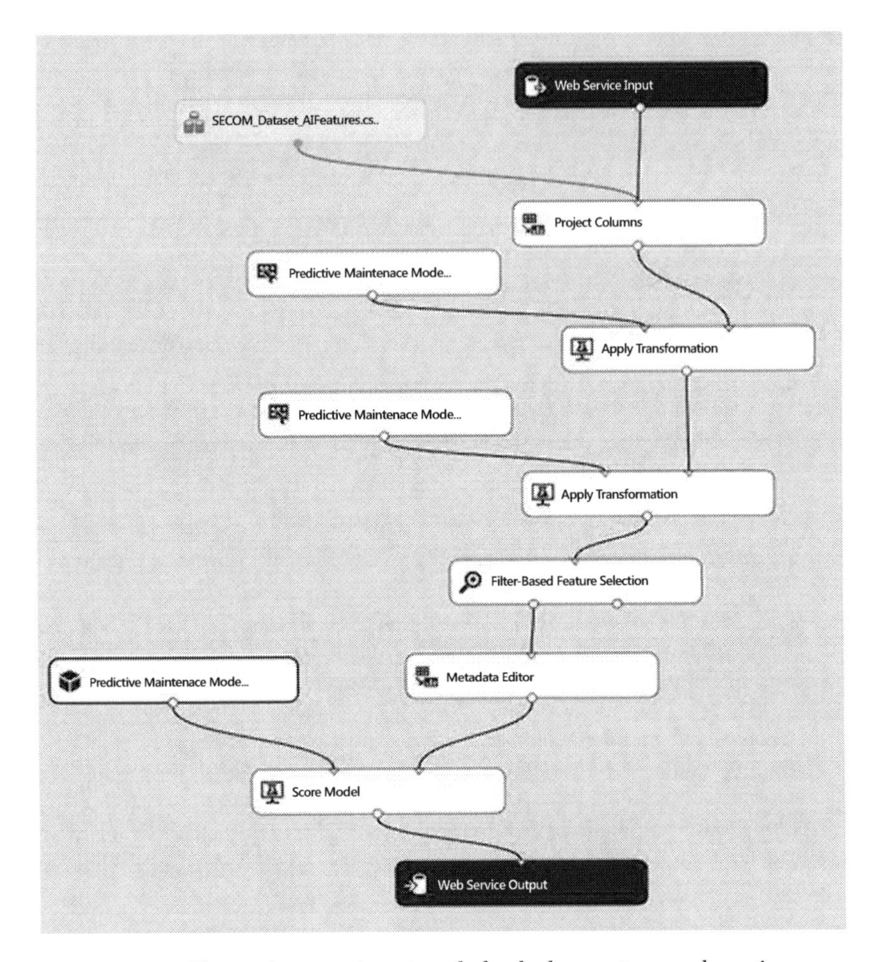

Figure 11-13. *The scoring experiment ready for deployment as a web service*

Publishing Your Experiment as a Web Service

At this point your model is now ready to be deployed as a web service. To do this, follow these steps.

1. Run your experiment with the **Run** button at the bottom of Azure Machine Learning Studio.
2. Click the **Deploy Web Service** button at the bottom of Azure Machine Learning Studio. The tool will deploy your scored model as a web service. The result should appear as shown in Figure 11-14.

Congratulations, you have just published your machine learning model into production. Figure 11-14 shows the API key for your model as well as the URLs you will use to call your model either interactively in request/response mode, or in batch mode. It also shows a link to a new Excel workbook you can download to your local file system. With this spreadsheet you can call your model to score data in Excel. See Chapter 8 for details on how to visualize your results with this Excel workbook. In addition, you also get sample code you can use to invoke your new web service in C#, Python, or R.

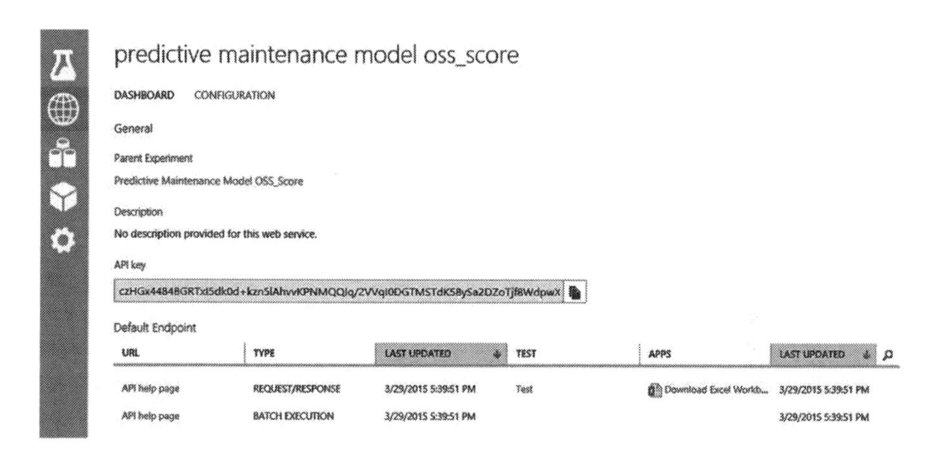

Figure 11-14. *The results of the deployed model showing the API key, URLs for the service, and a link to a new Excel workbook you can download*

Summary

You began this chapter by exploring predictive maintenance, which shows the business opportunity and the promise of machine learning. Using the SECOM semiconductor manufacturing data from the University of California at Irvine's Machine Learning Data Repository, you learned how to build and deploy a predictive maintenance solution in Azure Machine Learning. In this step-by-step guide, we explained how to perform data preprocessing and analysis, which are critical for understanding the data. With that understanding of the data, we used a two-class logistic regression and a two-class boosted decision tree algorithm to perform classification. You also saw how to evaluate the performance of your models and avoid over-fitting. We explored how to use upsampling and downsampling to significantly improve the performance of your predictive model. Finally, you saw how to deploy the finished model in production as a machine learning web service on Microsoft Azure.

CHAPTER 12

Recommendation Systems

Are you thinking of building a recommendation engine? Or are you wondering how recommendations at your favorite website work? Look no further. This chapter builds on the introduction in Chapter 5 with a practical guide on recommendation engines. We will show step by step how to build recommendation engines in Azure Machine Learning.

Overview

Retail experiences are being reimagined. Whether you are shopping for a new book on Amazon, shopping for a new food blender on Taobao, looking for a movie to watch from Netflix or Hulu, or listening to songs on Last.fm or Pandora, personalized experiences enable you to make better choices and find the item you need.

As you search and browse the items, you see sections on the website that point out what others have bought, or items that you might be interested in. As friends in your social network make purchases, and share them on Facebook or Google+, you click the Like button. When you make a purchase at retail outlets or websites, you are encouraged to rate the items (such as movies) that you have bought or rented. These pleasurable and personalized experiences encourage you to explore, spend time with each of the services, find what you need quickly, and eventually buy some of these items.

Many companies realized that they are sitting on a treasure trove of data that can provide rich information on how their customers and the items purchased or rated are interconnected. Recommendation systems enable these companies to sieve through this data and figure out how to tap the insights derived from this data to deliver personalized experiences for their customers.

Note In 2009, Netflix offered a $1M Grand Prize to teams that were able to achieve a prediction accuracy that is 10% better than their home-grown recommendation engine called Cinematch. Read about the exciting innovations made by various teams around the world as they competed to build the best recommender system using data provided by Netflix at `http://www.netflixprize.com/` and `http://en.wikipedia.org/wiki/Netflix_Prize`.

Recommendation Systems Approaches and Scenarios

Various techniques for building recommendation systems have been explored over the years. These include

- **Collaborative Filtering**: By using information about the "tastes" (preferences) of the users of a system, collaborative filtering enables the recommendation system to infer the items that a user will like based on similar items they have liked in the past. Item-to-item collaborative filtering has been used at Amazon.com as part of their item recommendation process. In addition, companies such as Last.fm, Facebook, and LinkedIn have also used collaborative filtering when recommending music, friends, or folks you might know.
- **Content-Based Filtering**: Inspired by techniques from information retrieval, content-based filtering relies on item description and users' preference (type of item that they like) to figure out the type of items a user might like. Feedback from users (such as the Like button) can be used to influence weights given to each of the features used. Content-based filtering has been used in various companies such as Rotten Tomatoes, Internet Movie Database, and Pandora Radio.
- **Hybrid Approaches**: To improve the effectiveness of recommendation systems, hybrid approaches that use both collaborative and content-based filtering have been used. Netflix uses a hybrid approach to provide movie recommendations. Netflix uses information on the watching and searching behavior of users to figure out recommended movies. In addition, Netflix also recommends movies that are similar (based on the tags and description of the movie that the user has watched or rated).

Recommendation systems are commonly used in the following scenarios:

- **Related items or items that you might be interested in**: When a user is browsing or searching for a specific item, the user sees the recommendation for related items (such as items that are frequently bought together, or items that are related).
- **Recommendations provided on a website as users interact with the different sections in the site**: For example, when a user lands on the main page of a site (like Yahoo), she sees a personalized page on what's trending, news, and advertisements. This enables people to find the information that they need, and for companies to provide relevant news and advertisements to their customers.

- **Recommend applications in an App Store**: Given the huge collection of applications in various app stores (Microsoft, Apple, and Google), a mobile user needs to find the apps that they need quickly. Apps recommendations enable a mobile user to find what they need based on their preferences. When the mobile user uses the app, she can rate the app. This provides a feedback loop, and helps to refine future recommendations to the mobile user.

■ **Note** How does a recommendation systems work? Find out at `http://blogs.technet.com/b/machinelearning/archive/2014/07/09/recommendations-everywhere.aspx`.

The Business Problem

Imagine that you are a data scientist and part of a team at a startup that is building an award-winning restaurant recommendation service, the goal of which is to make restaurant recommendations available to users anytime, anywhere on the planet. Users of the service download your app, and regardless of their location (perhaps they are visiting a new city), and the app provides recommendations on the best restaurants.

In order to build a personalized experience for the user, the mobile application provides a list of recommended restaurants based on his preferences and his current location. Whenever a user visits one of the restaurants, she uses the mobile application to rate it and provides a review of the restaurant. You are tasked with building a machine learning model to provide recommendations of highly-rated restaurants to the customer based on her preferences.

This is a practical problem for many service providers that offer a highly curated list of restaurants, shopping, and other points of interest. For example, when someone visits a new city, they often use Yelp to find the best restaurants. In other parts of the world, similar services include HungryGoWhere in Singapore and Dianping in China. The ability to provide a personalized experience and the highly relevant recommendations encourage users of the app to "stick" with the service.

■ **Note** See how recommendation systems are being used at JJ Food Service, one of the largest independent food delivery services in the United Kingdom (UK) at `http://blogs.technet.com/b/machinelearning/archive/2015/01/08/azure-ml-predicts-customers-shopping-lists-even-before-they-shop.aspx`.

Data Acquisition and Preparation

Let's explore the datasets that will be used as inputs for building the recommendation system.

The Dataset

For this exercise, you will use the dataset that is available in the Azure Machine Learning Studio. Let's start to explore the dataset that will be used to build the Recommendation model. Azure Machine Learning Studio provides several experiments that can be used to jumpstart your learning on how to build a recommendation system.

In Azure Machine Learning Studio, do the following:

1. Click **New Experiment**.
2. In the Search box, enter **Restaurant**.
3. You will see the experiments that are used for training recommenders (shown in Figure 12-1).

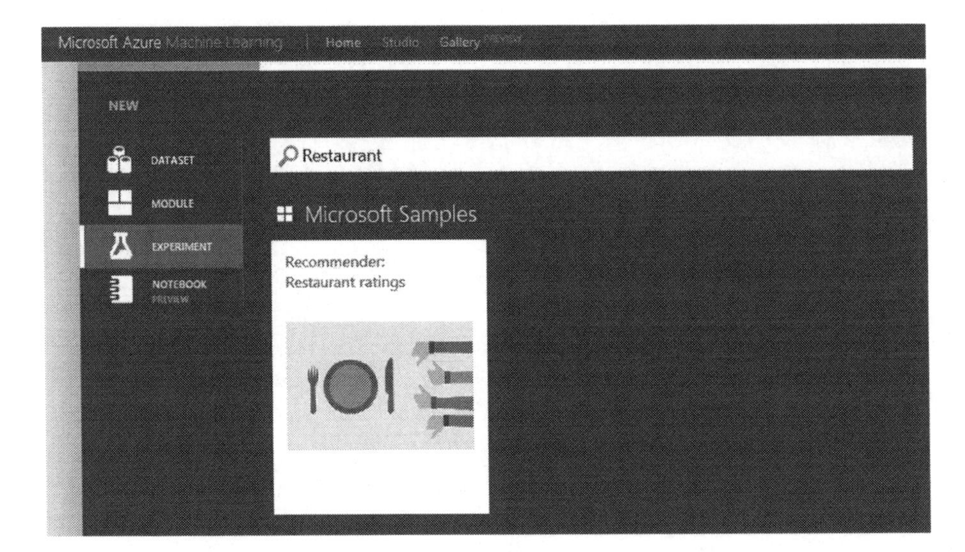

Figure 12-1. *Using the Restaurant Recommender*

4. Select the **Recommender: Restaurant ratings** by hovering on the image, and choose **Open in Studio** (shown in Figure 12-2).

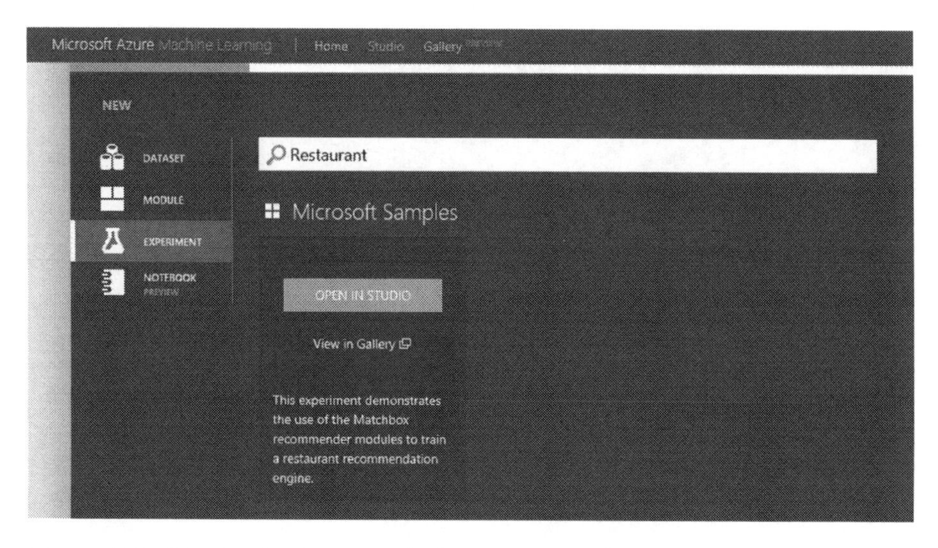

Figure 12-2. *Selecting the Recommender for restaurant rating*

After opening the experiment in Azure Machine Learning Studio, you will see the following datasets:

- **Restaurant Ratings**: Ratings made by each user for a restaurant
- **Restaurant Customer**: Customer profile, such as whether the customer is a smoker, how often the customer drinks alcohol, dress preference, preferred ambience, marital status, etc.
- **Restaurant Feature**: Restaurant information, such as the city and state in which the restaurant is located, whether the restaurant serves alcohol, dress code for the restaurant, etc.

Let's explore what's in each of the datasets.

1. Right-click the **output node** of the **Restaurant Ratings** dataset and choose **Visualize** (shown in Figure 12-3).

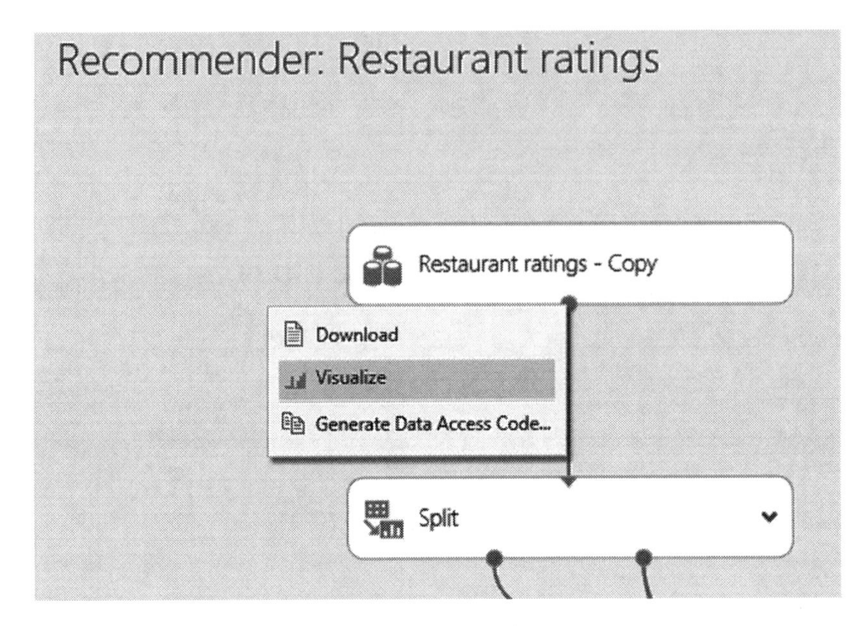

Figure 12-3. *Visualizing the Restaurant ratings dataset*

2. After clicking Visualize, you will see that the Restaurant Ratings dataset consists of three columns: `userID`, `placeID`, and `rating` (shown in Figure 12-4). These three columns correspond to the rating that has been made by a user at a specific restaurant. The `userID` and `placeID` are used to identify a specific user and restaurant, respectively.

Recommender: Restaurant ratings › Restaurant ratings (copy 1) › dataset

rows 1161 columns 3

userID	placeID	rating
U1077	135085	2
U1077	135038	2
U1077	132825	2
U1077	135060	1
U1068	135104	1
U1068	132740	0
U1068	132663	1
U1068	132732	0
U1068	132630	1

Figure 12-4. *Restaurant ratings dataset columns (userID, placeID, and rating)*

Let's dive deeper into the dataset to understand how the data is distributed and the number of ratings made by each user. To do that, do the following.

1. Click the `userID` column (shown in Figure 12-5).

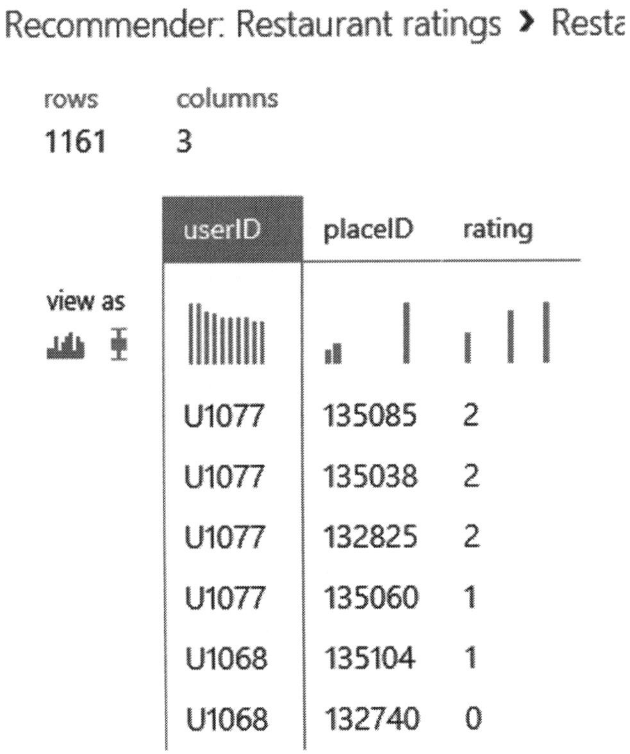

Recommender: Restaurant ratings › Restɛ

rows 1161
columns 3

view as

userID	placeID	rating
U1077	135085	2
U1077	135038	2
U1077	132825	2
U1077	135060	1
U1068	135104	1
U1068	132740	0

Figure 12-5. *Selecting userID to see the number of ratings made by a user*

2. In the **Visualization** pane on the right, you will see the number of ratings made by the first 10 users (shown in Figure 12-6).

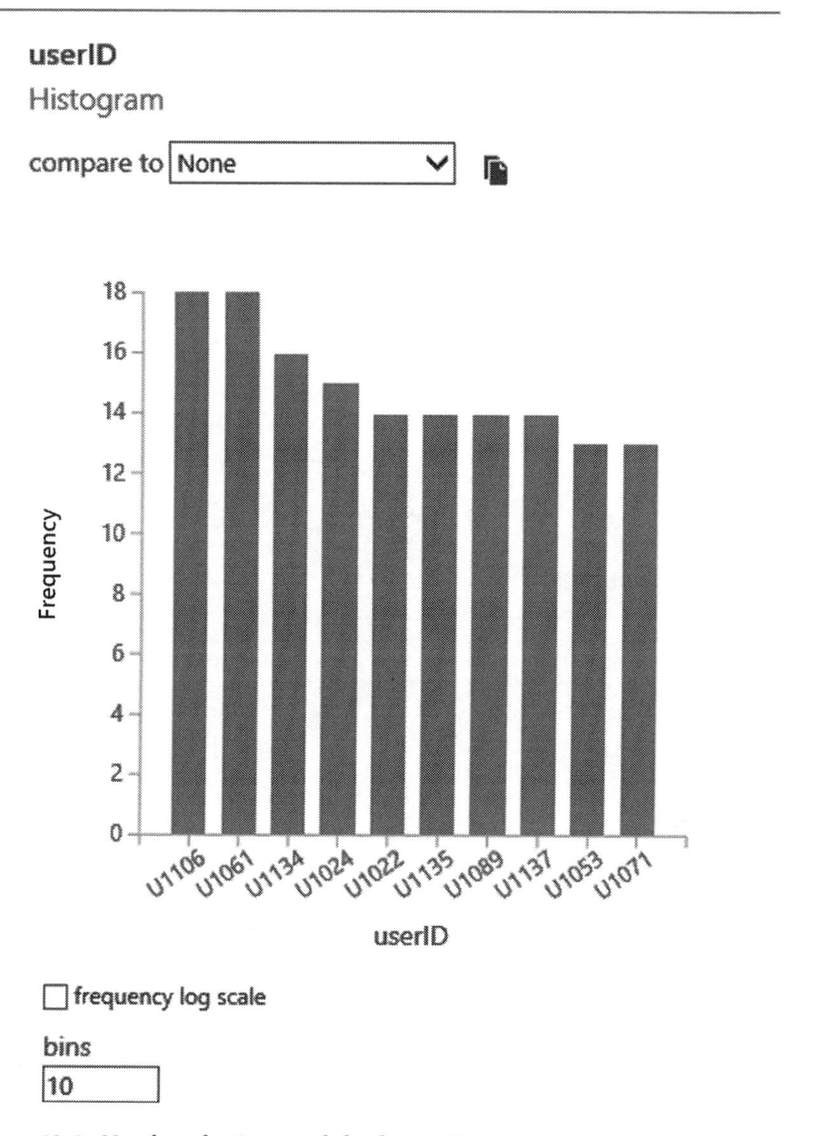

Figure 12-6. *Number of ratings made by the top 10 users*

Note that the default number of bins used is 10. To see more users, you can increase the number of bins.

Next, let's look at the restaurant customers and restaurant feature datasets. In the Restaurant Customer and Restaurant Feature dataset, you will see the columns `userID` and `placeID`, respectively. The `userID` and `placeID` columns are used as keys to uniquely identify a specific customer or restaurant, and are shown earlier in the Restaurant Ratings dataset (Figure 12-5).

1. Right-click the **output node** of the **Restaurant Customer** dataset and choose **Visualize** (shown in Figure 12-7).

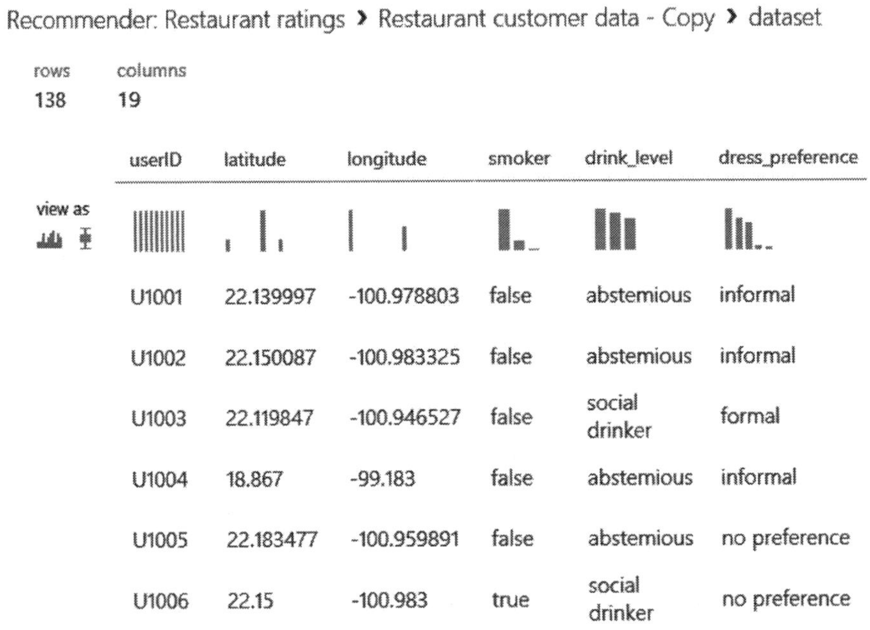

Recommender: Restaurant ratings › Restaurant customer data - Copy › dataset

rows	columns
138	19

userID	latitude	longitude	smoker	drink_level	dress_preference
U1001	22.139997	-100.978803	false	abstemious	informal
U1002	22.150087	-100.983325	false	abstemious	informal
U1003	22.119847	-100.946527	false	social drinker	formal
U1004	18.867	-99.183	false	abstemious	informal
U1005	22.183477	-100.959891	false	abstemious	no preference
U1006	22.15	-100.983	true	social drinker	no preference

***Figure 12-7.** Restaurant customer dataset*

2. Right-click the **output node** of the **Restaurant Feature** data set and choose **Visualize** (shown in Figure 12-8).

Recommender: Restaurant ratings › Restaurant feature data - Copy › dataset

rows 130 columns 21

placeID	latitude	longitude	the_geom_meter	name
134999	18.915421	-99.184871	0101000020957F000088568DE356715AC138C0A525FC464A41	Kiku Cuernavaca
132825	22.147392	-100.983092	0101000020957F00001AD016568C4858C1243261274BA54B41	Puesto de Tacos
135106	22.149709	-100.976093	0101000020957F0000649D6F21634858C119AE9BF528A34B41	El Rincon de San Francisco
132667	23.752697	-99.163359	0101000020957F00005D67BCDDED8157C1222A2DC8D84D4941	Little Pizza Emilio Portes Gil
132613	23.752904	-99.165076	0101000020957F00008EBA2D06DC8157C194E03B7B504E4941	Carnitas Mata
135040	22.135617	-100.969709	0101000020957F00001B552189B84A58C15A2AAEFD2CA24B41	Restaurant Los Compadres

Figure 12-8. *Restaurant feature dataset*

In the experiment, these three datasets will be used as inputs to the **Train Matchbox Recommender** module (shown in Figure 12-9). In Figure 12-9, you can see that the restaurant rating data is split into a training and test dataset using the **Split** module (below the Restaurant ratings - Copy dataset).

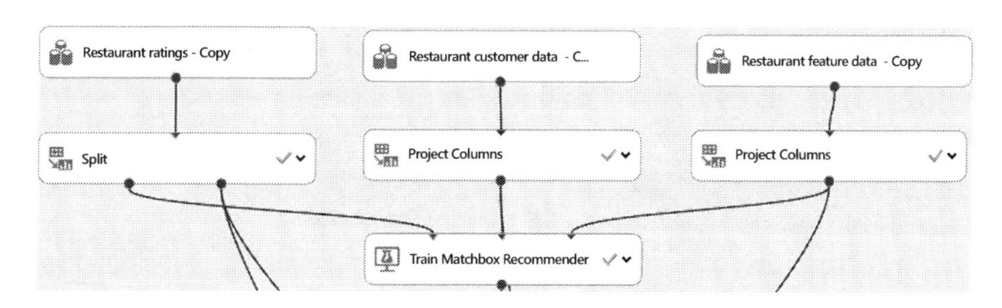

Figure 12-9. *Inputs to the Train Mathbox Recommender module*

Click the **Split** module, and you will see a Recommender Split instead of the default Split (Split Rows). Figure 12-10 shows the settings for the Split module. The Recommender Split is specially designed for recommender systems. It ensure that user-item pairs (and ratings) are evenly divided between the training and test datasets. The Recommender Split provides several properties that can be configured by the user.

- **Fraction of training-only users**: Specifies the fraction of users that will be used as part of training.
- **Fraction of test user ratings for training**: Specifies the fraction of user ratings that will be used as part of training.
- **Fraction of cold users**: Cold users refer to users where no user information is available. Using cold users in an experiment is useful because it enables the recommendation system to make recommendations for users where you do not have complete information.
- **Fraction of cold items**: Cold items refer to items where no item information is available. In this experiment, this refers to restaurants where you do not have complete information.
- **Fraction of ignored users**: This is commonly used for performance tuning, where you want the recommendation system to be trained quickly using a subset of the users available.
- **Fraction of ignored items**: This is commonly used for performance tuning, where you want the recommendation system to be trained quickly using a subset of the items available.
- **Remove occasionally produced cold items**: This is used to ensure that all users and items in the test dataset are included as part of the training dataset (when the fraction of cold users and items are set to zero).

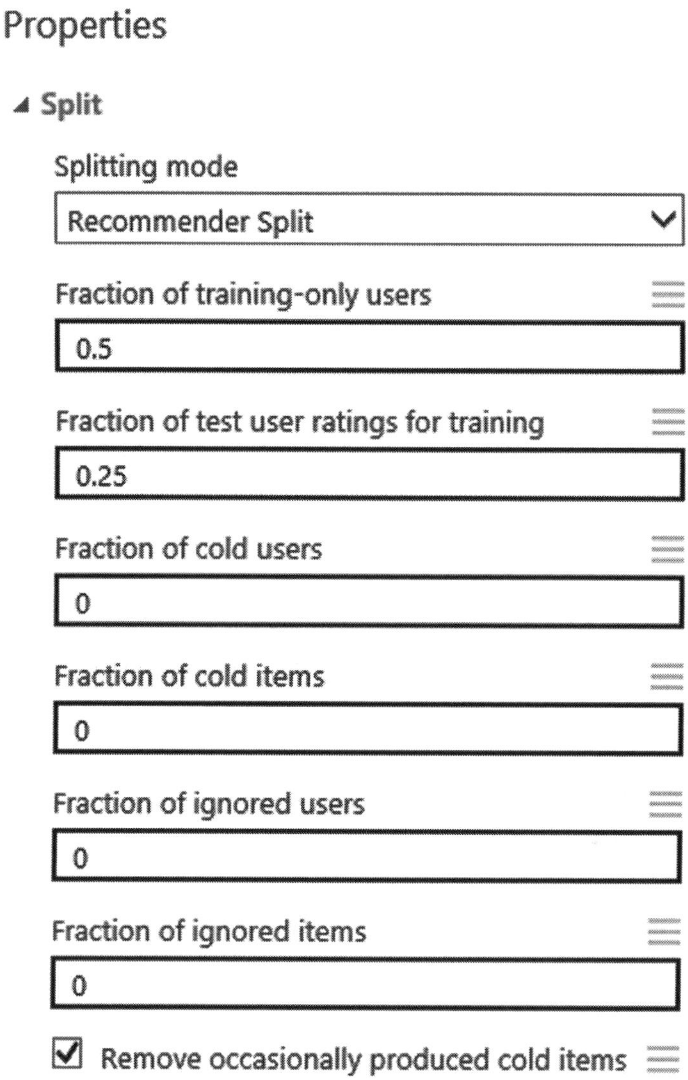

Properties

◢ Split

Splitting mode

Recommender Split

Fraction of training-only users

0.5

Fraction of test user ratings for training

0.25

Fraction of cold users

0

Fraction of cold items

0

Fraction of ignored users

0

Fraction of ignored items

0

☑ Remove occasionally produced cold items

Figure 12-10. *Using the Recommender Split*

Training the Model

Having completed the data preprocessing, the next step is to train the Matchbox Recommender using the **Train Matchbox Recommender Module**.

The goal of recommendation systems is to provide recommendations of one or more items to users of the system. In this experiment, the system recommends restaurants to the users of the system. Often, when a user is new to the system (when the user has not made any ratings for restaurants), recommendations are made using information about the user. This enables the Matchbox recommender to address a common issue faced by most recommendation systems, referred to as "cold-start." Over time, as the user uses the system and provides more ratings, the Matchbox recommender will be able to use past ratings made by the user to make recommendations.

■ **Note** Matchbox is a large-scale Bayesian Recommender System developed by Microsoft Research. It is a hybrid recommender that combines the strength of collaborative and content-based filtering. The Matchbox recommender algorithm is an online learning algorithm that enables the system to incrementally use new data to make sure the system is updated with the latest user preference. Refer to `http://research.microsoft.com/en-us/projects/matchbox` to learn more about the Matchbox Recommendation System.

The **Train Matchbox Recommender** module (Figure 12-11) requires three inputs:

- **User-item-rating triplet**: Rating made by a specific user for an item
- **Users feature**: Information about the user
- **Items feature**: Information about the item

Figure 12-11. *Train Matchbox Recommender module*

After you have connected the three input datasets to the Train Matchbox Recommender, click the Train Matchbox Recommender module to learn about the two properties that can be set by the user:

- **Number of traits**: This determines the number of traits that are learned for each user and item. The default value is 4. This can be fine-tuned to a higher number, which will increase the accuracy of the predictions made by the Matchbox Recommender. Usually, this value is set to be between 2 and 20.
- **Number of recommendation algorithm iterations**: This determines the number of times the Matchbox recommender algorithm iterates over the training data. The default value is 20.

When the value for either of these properties is increased, the training time also increases. The risk of overfitting increases too. Hence, you need to carefully balance how accurate you need the recommendation systems to be, how much time you are willing to spend on training the system, and how much overfitting you will accept. You are now ready to run the experiment. Click **Run** to run the experiment. Figure 12-12 shows the completed experiment.

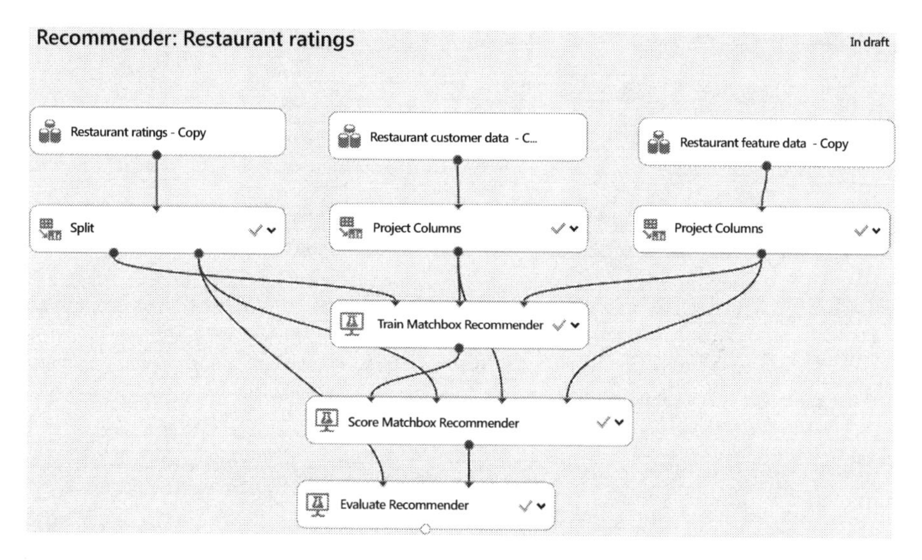

Figure 12-12. *Recommender - Restaurant ratings experiment*

■ **Note** Various recommendation web services are also available in the Azure Marketplace to help you get started quickly.

Recommendations: The recommendation web service helps provide item recommendation. See `http://datamarket.azure.com/dataset/amla/recommendations`.

Model Testing and Validation

After you have successfully run the experiment, you can see the results of the scoring by clicking the Score Matchbox Recommender module. To do this, follow these steps.

1. Select the **output node** of the **Score Matchbox Recommender** module, and click **Visualize** (shown in Figure 12-13).

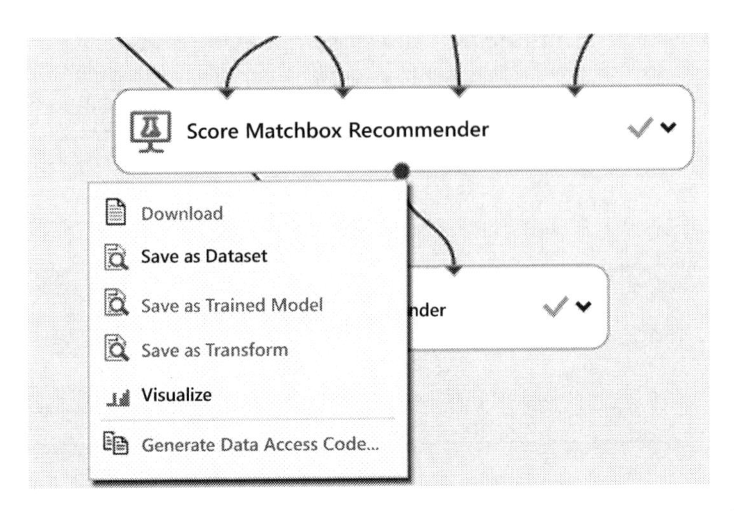

***Figure 12-13.** Visualizing the scored dataset of a Matchbox Recommender*

2. After you click Visualize, you will see the restaurant recommendations made for each of the user (shown in Figure 12-14).

Recommender: Restaurant ratings › Score Matchbox Recommender › Scored dataset

rows 69 columns 6

User	Item 1	Item 2	Item 3	Item 4	Item 5
U1048	135026	135034	135065	132723	135049
U1117	135018	135088			
U1049	135052	132862	135032	135085	135051
U1088	135057	135071	135032	135070	135108
U1062	135052	135045	135062	135085	135038
U1035	134986	135018	132583		
U1125	135062	135076	135042	135038	135032
U1013	135075	135079	132921	135072	135076
U1042	134986	132768	135021	134992	
U1123	132584	132667	132733	132608	132594
U1086	135045	132922	135085	132921	132937

***Figure 12-14.** Scored dataset for the Matchbox Recommender*

3. For each of the user, you will see that up to five recommendations are made. For example, user U1048 (first row) has the restaurant recommendations (identified by placeID) 135026, 135034, 135065, 132723, and 135049.

4. You can configure the number of item recommendations made by configuring the **Score Matchbox Recommender** module (shown in Figure 12-15). There are four properties that you can configure. These include

   a. **Recommender prediction kind**: This determines whether you require Item Recommendation, Related Users, Related Items, or Ratings Prediction.

   b. **Recommended item selection**: From Rated Items, or for all items.

   c. **Maximum number of items to recommend to a user**: The default value is set to 5.

   d. **Minimum size of the recommendation pool for single user**: The default value is set to 2. This determines the minimum number of items for the item that is being recommended.

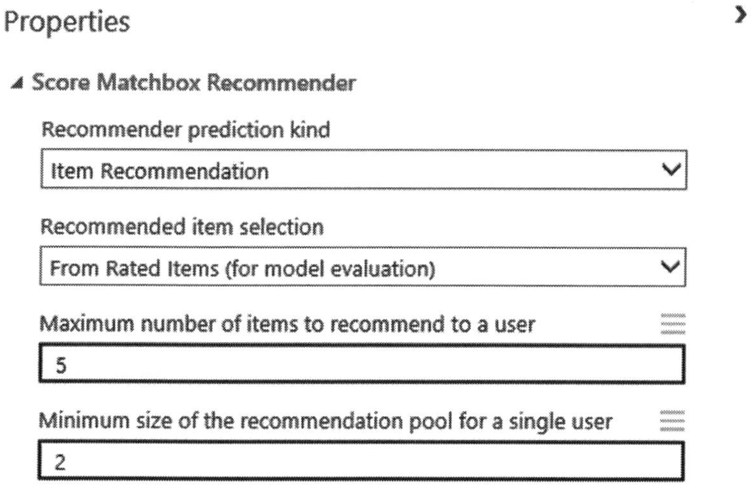

Figure 12-15. *Properties of Score Matchbox Recommender*

The Matchbox Recommender enables you to deliver the results for different prediction types. You can try changing the Recommender Prediction Kind to **Related Users** or **Related Items**. After you change the prediction kind, run the experiment. Click **Visualize** on the **Score Matchbox Recommender** module to see the results. Figure 12-16 shows the related users for each user, and Figure 12-17 shows the related items for each item.

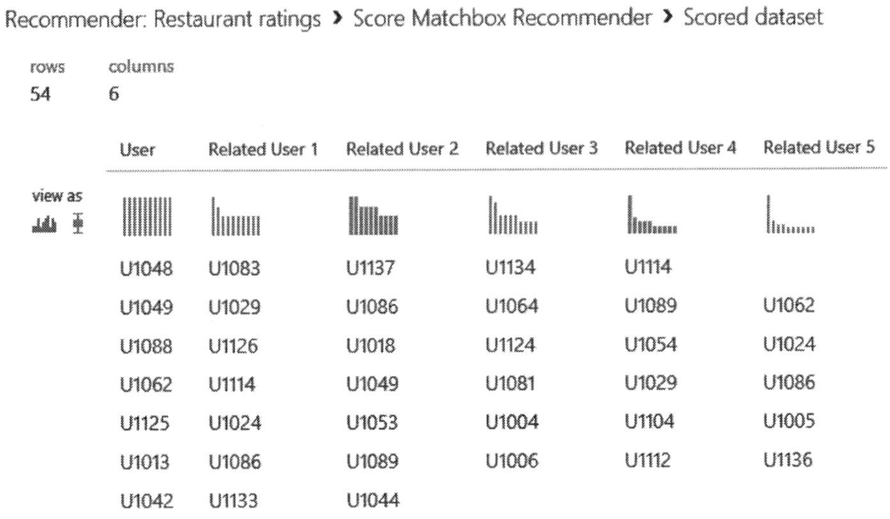

Recommender: Restaurant ratings › Score Matchbox Recommender › Scored dataset

rows 54 columns 6

User	Related User 1	Related User 2	Related User 3	Related User 4	Related User 5
U1048	U1083	U1137	U1134	U1114	
U1049	U1029	U1086	U1064	U1089	U1062
U1088	U1126	U1018	U1124	U1054	U1024
U1062	U1114	U1049	U1081	U1029	U1086
U1125	U1024	U1053	U1004	U1104	U1005
U1013	U1086	U1089	U1006	U1112	U1136
U1042	U1133	U1044			

Figure 12-16. *Related Users predictions*

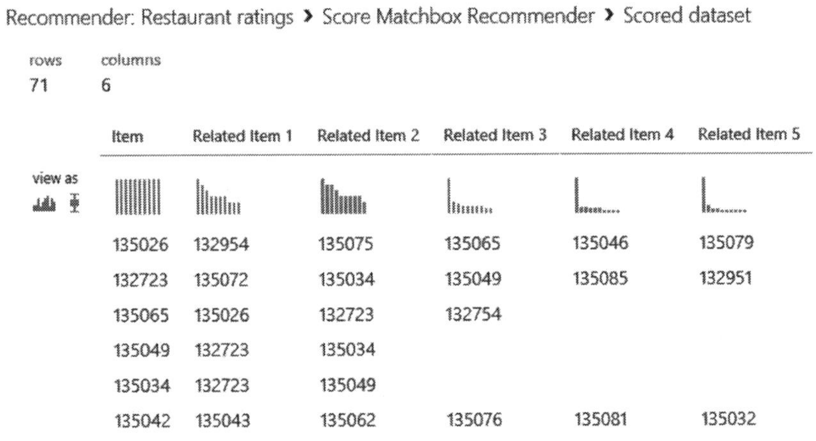

Recommender: Restaurant ratings › Score Matchbox Recommender › Scored dataset

rows 71 columns 6

Item	Related Item 1	Related Item 2	Related Item 3	Related Item 4	Related Item 5
135026	132954	135075	135065	135046	135079
132723	135072	135034	135049	135085	132951
135065	135026	132723	132754		
135049	132723	135034			
135034	132723	135049			
135042	135043	135062	135076	135081	135032

Figure 12-17. *Related Items predictions*

After running the experiment, you can click the **Evaluate Recommender** module. The output of the Evaluate Recommender module is dependent on the prediction kind that is selected, as follows:

- **Item Recommendation**: The computed normalized discounted cumulative gain (NDCG) is used to validate the accuracy of the items recommended. The gain is computed using the ratings from the test dataset. In this experiment, you can see the NDCG for item recommendations (shown in Figure 12-18).

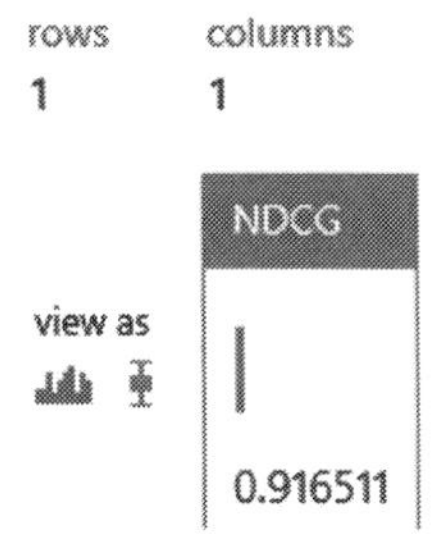

Figure 12-18. *NDCG for Item Recommendations*

- **Rating Prediction**: Two values are used to determine the accuracy of the rating prediction: mean absolute error (MAE) and root mean squared error (RMSE). If you change the Recommender Prediction Kind to Rating Prediction, you can see these two values (shown in Figure 12-19).

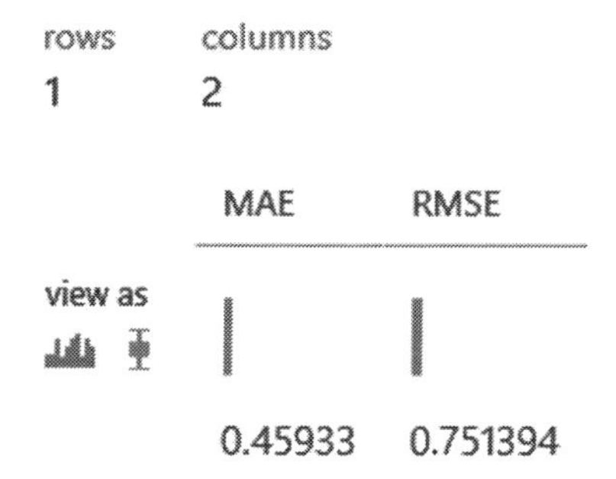

Figure 12-19. *MAE and RMSE for Rating Prediction*

- **Related Users and Related Items**: Two values are used to determine the accuracy of the related users prediction: L1 SIM NDCG using Manhattan distance is used to measure similarity, and L2 Sim NDCG uses Euclidean distance to measure similarity.

■ **Note** The Discounted Cumulative Gain (DCG) is commonly used in information retrieval to measure ranking quality. DCG measures the gain of an item based on its position in the result list.

The Normalized Discounted Cumulative Gain (NDCG) is used to ensure that the position of items is normalized, with values ranging from 0.0 to 1.0.

See the Technical Notes at `https://msdn.microsoft.com/en-us/library/azure/dn905954.aspx` to understand how the NDCG value is computed.

Various datasets are available for Recommendation System experiments:

MovieLens Dataset: Three MovieLens datasets are available for experiments. The datasets contain 100,000, 1 million, 10 million, and 20 million ratings collected from the `movielens.org` website. The MovieLens website provide non-commercial, personalized movie recommendations, and is run by the GroupLens research group of the University of Minnesota. See `http://grouplens.org/datasets/movielens/`.

Entree Chicago Recommendation Dataset: The dataset contains users interaction with the Entrée Chicago restaurant recommendation system over a three year period. See `http://archive.ics.uci.edu/ml/datasets/Entree+Chicago+Recommendation+Data`.

Summary

In this chapter, you learned about various recommendation system scenarios. This list of scenarios is non-exhaustive, and the ability to harness the power of recommendation systems to deliver personalized experiences enables many companies to delight their customers, drive revenue growth, and inspire customers to "stick" with the services provided.

We shared with you the different types of recommendation systems and the problem space addressed by each type. Using the Restaurant Ratings Recommender example from the Azure Machine Learning gallery, you learned how to harness the capabilities of the Matchbox Recommender to predict various recommendation kinds, and how to interpret the metrics used to measure the effectiveness of the recommender.

CHAPTER 13

Consuming and Publishing Models on Azure Marketplace

The Azure Machine Learning Studio makes it easy to create new predictive models. But what if there was a way to harness the power of machine learning without having to understand the data science behind it? This is the promise of the Azure Machine Learning Marketplace. Think of it as the place where you can go find "baked" APIs that solve interesting problems thanks to machine learning. This chapter will introduce this marketplace, showing existing solutions from Microsoft and its partners. We will also show step by step how you can sell your own predictive models on Azure Marketplace.

What Are Machine Learning APIs?

The Machine Learning APIs enable developers or organizations to consume machine learning solutions instead of creating them. This is particularly important since the supply of good data scientists is very low at the moment. With these APIs, organizations can leverage the power of machine learning without hiring data scientists. The marketplace also allows data scientists to amplify their reach by selling their solutions to many more customers.

These APIs are exposed to the public in several locations today. They are exposed in the Azure Machine Learning Gallery, as shown in Figure 13-1. They can also be found in the Microsoft Azure Marketplace. The Microsoft Azure Marketplace is an online store for thousands of applications, add-ons, APIs, and data that are preconfigured to run on Windows Azure. Azure Marketplace does not sell only Microsoft products; it also has offerings from many vendors, including competitors such as Oracle, Box, MungoBD, Cloudera, and others. The store has many categories of products for virtual machines, application services, web applications, and many more. In this book, we will only focus on machine learning APIs.

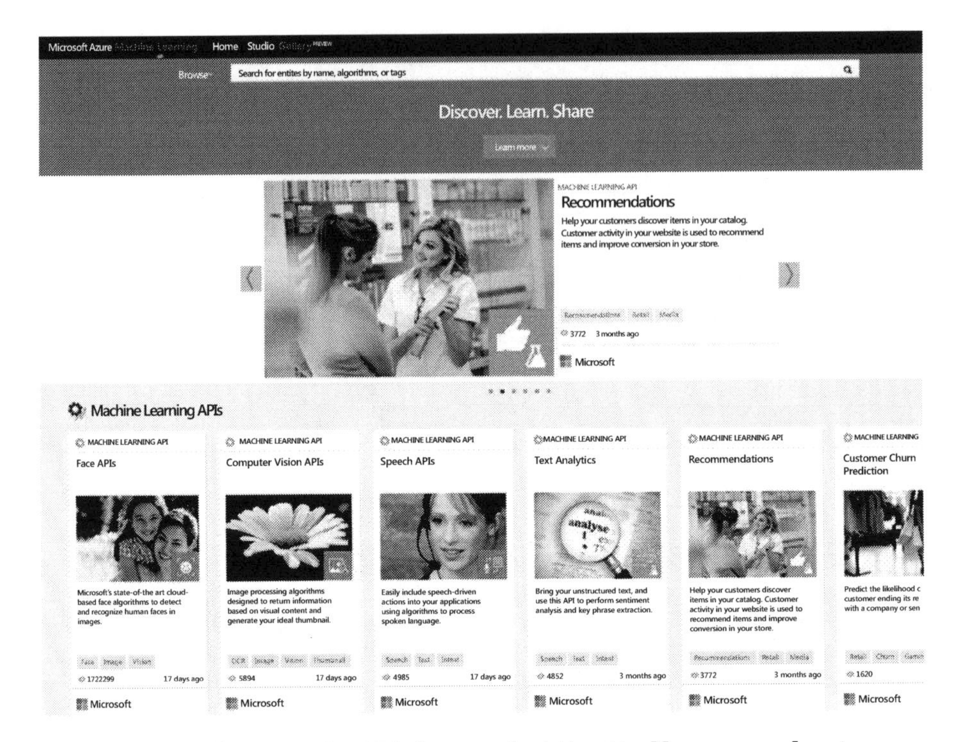

Figure 13-1. *Machine Learning APIs showcased at* `http://gallery.azureml.net`

As you can see in Figure 13-1, there are many ready-to-consume APIs, including the following:

- **Face APIs**: These APIs provide state-of-the-art algorithms to process face images, like face detection with gender and age prediction, face recognition, and face alignment. It is based on Bing technology. For more details, see `www.projectoxford.ai/face`.
- **Speech APIs**: Speech APIs provide the ability to process spoken language. With these APIs, developers can easily include the ability to convert spoken audio to text and convert text to speech. In certain cases, the APIs also allow for real-time interaction with the user as well. For more information, see `www.projectoxford.ai/speech`.
- **Computer Vision APIs**: The Computer Vision APIs are a collection of state-of-the-art image processing algorithms. For instance, you can analyze an image to categorize it, understand if it is adult content, or get the dominant color of the image. The API also supports extraction of text from imagery using Optical Character Recognition (OCR), and has a thumbnail generation capability. For more information, see `www.projectoxford.ai/vision`.

- **Recommendations:** This is a recommendation engine that allows you to do cross-selling or upselling on your website. Provided by Microsoft, this API offers both item-to-item and user-item recommendations. With item-to-item recommendations you can make suggestions based on the fact that those who bought product A also bought product B. For more personalized recommendations you can use the user-to-item recommendations based on the fact that users like you bought product A. Full details on this service is at `https://datamarket.azure.com/dataset/amla/recommendations`.
- **Text Analytics:** This API implements text mining using Azure Machine Learning. It uses Natural Language Processing techniques to mine data in English only, and no training is required. You can use it for sentiment analysis: in this case the API returns a score between zero and one. A score close to one indicates positive sentiment, while scores near zero denote negative sentiment. More details are available at `https://datamarket.azure.com/dataset/amla/text-analytics`.
- **Customer Churn Prediction**: Built by Microsoft with Azure Machine Learning, this API predicts the likelihood of attrition; it identifies which customers are most likely to terminate their relationship with their providers or retailers. More details are available at `https://churn.cloudapp.net/`.
- **Forecasting with Autoregressive Integrated Moving Average (ARIMA):** You can use this API to build forecasting models through the ARIMA time series algorithm. With this API you can address problems such as demand forecasting to forecast how much electricity will be consumed on a given week, or how many products will be sold in a given day. More details on this API are available at `https://datamarket.azure.com/dataset/aml_labs/arima`. Note that Azure Marketplace also offers two other APIs that use different Time Series algorithms. The first uses Exponential Smoothing (ETS), while the second employs both ETS and Seasonal Trend Decomposition (STL) to predict future outputs based on historical data. These are all built with Azure Machine Learning and published by Microsoft.
- **Giving Score:** Developed by Versium Analytics, this API enables nonprofit organizations to determine which of their contributors are most likely to make large donations or become repeat contributors. They can also use it to identify which prospects have a high probability to donate to charity. More details can be obtained at `https://datamarket.azure.com/dataset/versium/lifedata-giving-score`.

How to Use an API from Azure Marketplace

In this section, you will see how to use machine learning models from the Azure Marketplace. Specifically, you will see how to use the Recommendations API that was introduced in the previous section. We will demonstrate the Recommendations API with a sample application that is available on Azure Marketplace.

The goal of the Recommendations API is to generate a model that will use prior transaction data to recommend items that are of interest to customers. For instance, the recommendations model should be able to suggest to your customers which items they are most likely to want to purchase once they visit a particular product page. It can also recommend to the customer which other products they may want to visit, as shown in Figure 13-2.

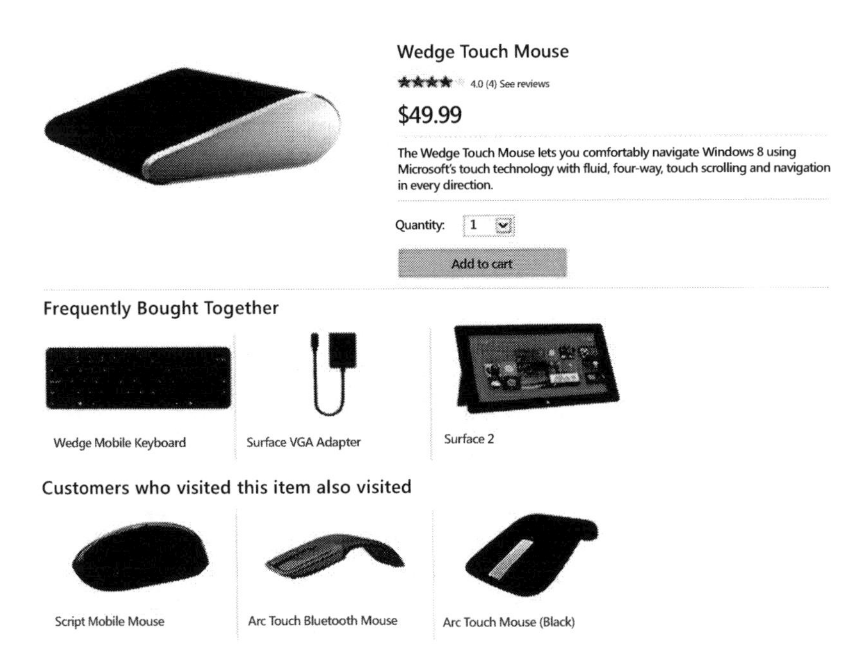

Figure 13-2. *Sample Usage of the Recommendations API*

To provide a concrete example from retail, the Recommendations API will allow you create a Customized Recommendations Model where, given an item (or a set of items) that the customer is interested in, it will predict which other products the customer is likely to want to purchase.

As outlined in Figure 13-3, in order to generate this predictive model, the Recommendations API needs you to provide two pieces of data:

- **Your catalog**: In the retail example, this is data about the products you sell. In its simplest format, you can think of this file as rows with information with the following format:

```
<Item Id>,<Item Name>,<Item Category>[,<Description>]
```

- **Your previous transactions**: This is also called usage data. This file should have rows that describe prior user behavior so that the model can be trained. The schema for this file can be as simple as <User Id>,<Item Id>[,<Time>,<Event>]

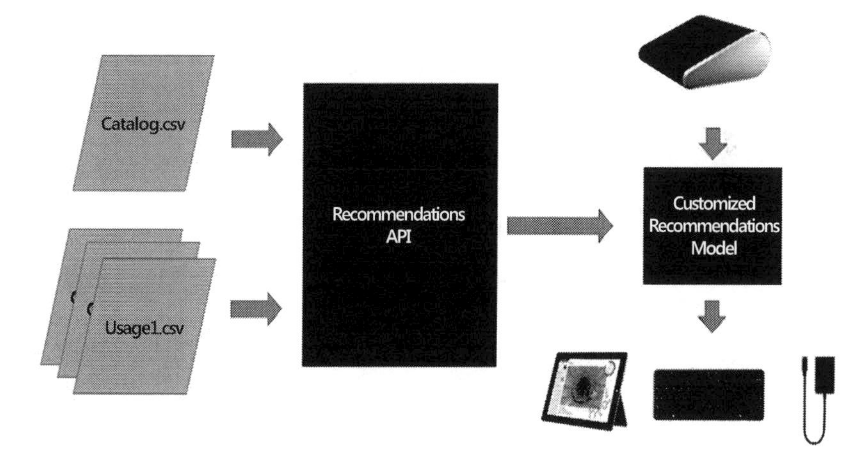

Figure 13-3. Recommendations API Diagram

Once this data is fed into the Recommendations API, it can create your customized recommendations model. Figure 13-3 shows the outline of the recommendation API.

You can obtain the sample application from the Recommendations API page as follows:

1. Visit the Recommendations API page at `https://datamarket.azure.com/dataset/amla/recommendations`.
2. The right pane shows pricing information. Under the pricing section find the link named **Sample Application Guide**. This link has information on the sample application. This right pane also has a link for **Sample Application**. Click this link to download the sample application.
3. Now download the sample application on your computer and extract the ZIP file.

■ **Note** To run the sample application, you must first sign up for the Recommendations API. To do this, go to the Recommendations API page at `https://datamarket.azure.com/dataset/amla/recommendations`. Sign up for the $0 subscription that provides up to 10,000 transactions per month for free!

Second, you also need an integrated development environment (IDE) for C# code. We use Visual Studio in this chapter.

After extracting the files from the sample application ZIP file, you can open the project file named **AzureMLRecoSampleApp.csproj** in Visual Studio or your preferred IDE. Follow these steps to build and test a recommender with this sample application.

1. Figure 13-4 shows the sample application (**SampleApp.cs**) in Visual Studio. Note that lines 37 and 38 read the username and account key from the arguments. So you need to provide your username and account key for Azure Marketplace. If you have not subscribed to Azure Marketplace, you may do so now as follows.

    a. Visit `https://datamarket.azure.com/`.

    b. Click the link for **Free Trial** at the top right of the page. Follow the instructions to sign up for a free trial.

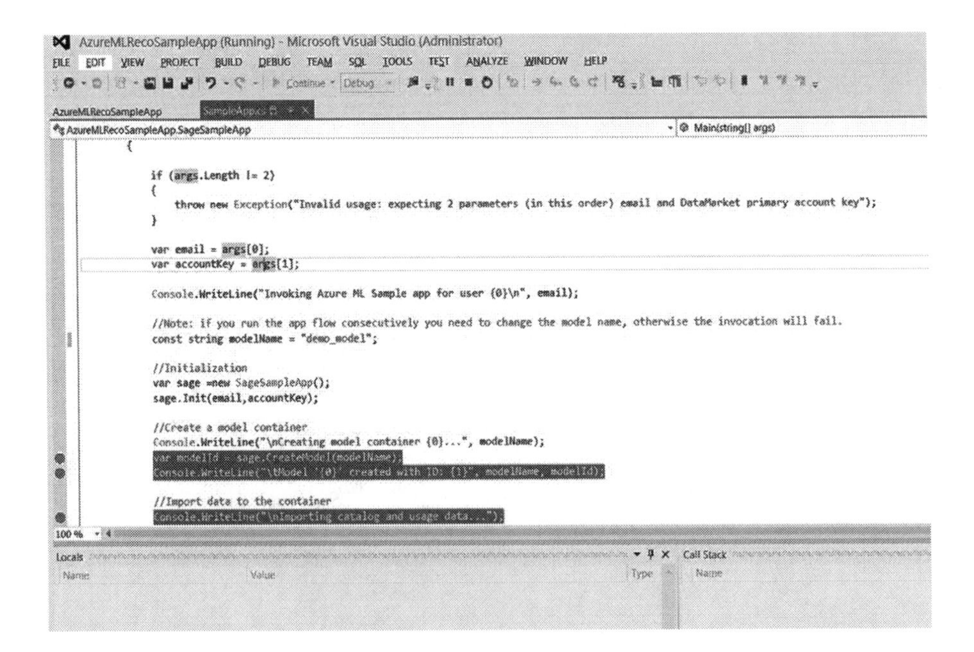

Figure 13-4. *Illustrating the sample application in Visual Studio*

2. Next, open the properties file called **AzureMLRecoSampleApp** in Visual Studio.
3. On line 58, the application loads the catalog from the sample catalog named **catalog_small.txt** in the **Resources** folder.
4. Similarly, line 62 reads usage data from the **usage_small.txt** file from the **Resources** folder.
5. Click the **Debug** tab on the left pane, and enter your username and account key in the text box named **command line arguments**. Note that you do not need commas between your arguments; simply leave spaces between your username and account key.
6. Now hit **F5** to compile and run the sample application. The results will appear on the console as show in Figures 13-5 and 13-6.

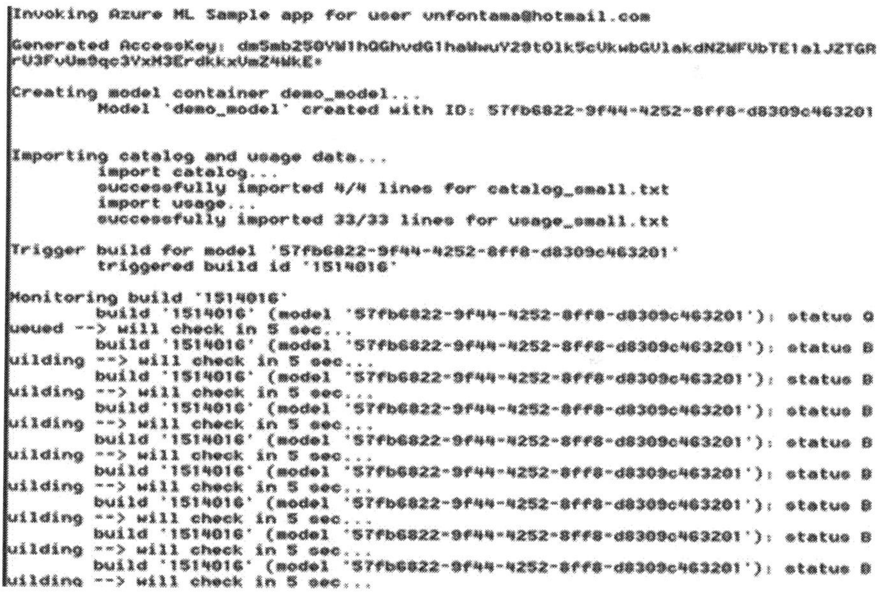

```
Invoking Azure ML Sample app for user unfontama@hotmail.com

Generated AccessKey: dm5mb25OVW1hQGhvdG1haWwuY29tO1k5cUkwbGU1akdNZWFUbTE1a1JZTGR
rU3FvUm9qc3VxM3ErdkkxUmZ4WkE=

Creating model container demo_model...
        Model 'demo_model' created with ID: 57fb6822-9f44-4252-8ff8-d8309c463201


Importing catalog and usage data...
        import catalog...
        successfully imported 4/4 lines for catalog_small.txt
        import usage...
        successfully imported 33/33 lines for usage_small.txt

Trigger build for model '57fb6822-9f44-4252-8ff8-d8309c463201'
        triggered build id '1514016'

Monitoring build '1514016'
        build '1514016' (model '57fb6822-9f44-4252-8ff8-d8309c463201'): status Q
ueued --> will check in 5 sec...
        build '1514016' (model '57fb6822-9f44-4252-8ff8-d8309c463201'): status B
uilding --> will check in 5 sec...
        build '1514016' (model '57fb6822-9f44-4252-8ff8-d8309c463201'): status B
uilding --> will check in 5 sec...
        build '1514016' (model '57fb6822-9f44-4252-8ff8-d8309c463201'): status B
uilding --> will check in 5 sec...
        build '1514016' (model '57fb6822-9f44-4252-8ff8-d8309c463201'): status B
uilding --> will check in 5 sec...
        build '1514016' (model '57fb6822-9f44-4252-8ff8-d8309c463201'): status B
uilding --> will check in 5 sec...
        build '1514016' (model '57fb6822-9f44-4252-8ff8-d8309c463201'): status B
uilding --> will check in 5 sec...
        build '1514016' (model '57fb6822-9f44-4252-8ff8-d8309c463201'): status B
uilding --> will check in 5 sec...
        build '1514016' (model '57fb6822-9f44-4252-8ff8-d8309c463201'): status B
uilding --> will check in 5 sec...
```

Figure 13-5. *Output of the application while the recommendation model is being built*

```
        Build 1514016 ended with status Success

Jpdating model description to 'book model' and set active build

Waiting for propagation of the built model...

Getting some recommendations...
         for single seed item
Recommendation for 'Id: 2406e770-769c-4189-89de-1c9283f93a96, Name: Clara Callan
'
           Name: Spadework, Id: 3BB5CB44-D143-4BDD-A55C-443964BF4B23, Rating: 0.5
, Reasoning: Most popular item (default system recommendation)
           Name: Restraint of Beasts, Id: 552A1940-21E4-4399-82BB-594B46D7ED54, R
ating: 0.5, Reasoning: Most popular item (default system recommendation)


Recommendation for 'Id: 552a1940-21e4-4399-82bb-594b46d7ed54, Name: Restraint of
 Beasts'
           Name: Spadework, Id: 3BB5CB44-D143-4BDD-A55C-443964BF4B23, Rating: 0.4
99994681837465, Reasoning: Most popular item (default system recommendation)



         for a set of seed item
        Recommendations for [Id: 2406e770-769c-4189-89de-1c9283f93a96, Name: Cla
ra Callan],[Id: 552a1940-21e4-4399-82bb-594b46d7ed54, Name: Restraint of Beasts]

           Name: Spadework, Id: 3BB5CB44-D143-4BDD-A55C-443964BF4B23, Rating: 0.4
99994681837465, Reasoning: Most popular item (default system recommendation)
Press any key to end
```

Figure 13-6. *Output of the application showing recommendations from the recommendation model*

Now let's see how the sample application works since it shows how you can call the Recommender API from your own C# application. Lines 32 to 47 in the `Main()` function initialize the SampleApp application with your username and account key. On line 51, the application creates a new container for the recommendation model with the following code:

```
var modelId = sage.CreateModel(modelName);
```

Lines 55 to 63 load the catalog and user data with the following code:

```
Console.WriteLine("\nImporting catalog and usage data...");
var resourcesDir = Path.Combine(Path.GetDirectoryName(Assembly.
GetExecutingAssembly().Location),"Resources");

Console.WriteLine("\timport catalog...");

var report =
sage.ImportFile(modelId, Path.Combine(resourcesDir,"catalog_small.txt"),
Uris.ImportCatalog);

Console.WriteLine("\t{0}", report);

Console.WriteLine("\timport usage...");
```

```
report = sage.ImportFile(modelId, Path.Combine(resourcesDir, "usage_small.txt"),
        ris.ImportUsage);

Console.WriteLine("\t{0}", report);
```

On line 68 the application triggers a build to create a recommendation model with the following code:

```
var buildId = sage.BuildModel(modelId,"build of " + DateTime.UtcNow.
ToString("yyyyMMddHHmmss"));
```

Full details of the `BuildModel` function are on lines 198 to 232. The `while` loop on lines 76 to 90 checks the status of the model building. It checks the build status every 5 seconds until the job completes. Figure 13-5 shows the output of the application on the console.

When the model is built successfully, you can start testing with a few examples. In this case, the sample application tests the model with two examples from the catalog file (namely the books *Clara Callan* and *Restraint of Beasts*). On line 120 the application uses the following function call to get recommendations based on a single item:

```
sage.InvokeRecommendations(modelId, seedItems,false);
```

In this case, the sample application asks for a recommendation based on the book *Clara Callan*. In response, the model recommends two books, *Spadework* and *Restraint of Beasts*.

Line 125 also calls the **InvokeRecommendations** function. This time it passes true as the last function parameter. Setting this last argument to true allows you to get recommendations using many items as the seed. In this case, the sample application uses the books *Clara Callan* and *Restraint of Beasts* as the seed. For this set of items, the recommendation model returns one item, the book *Spadework*. The results are shown in the output listing in Figure 13-6.

```
sage.InvokeRecommendations(modelId, seedItems, true);
```

More details of the function `InvokeRecommendations` are available in lines 366 to 390 of the sample application. It is worth noting that you cannot build the sample application more than once with the same model name. On line 43 the model name is set to the constant string `"demo_model"`. To rerun the sample application you need to change the name of the model in line 43; otherwise an exception will be thrown.

```
const string modelName = "demo_model";
```

Publishing Your Own Models in Azure Marketplace

In this section, you will see how to publish machine learning models in Azure Marketplace using the predictive maintenance model from Chapter 11 as an example. You can publish your models in Azure Marketplace with this three-step process.

1. Create and publish a web service for your machine learning model.
2. Obtain its API key and the details of the OData endpoint.
3. Publish your model as an API in Azure Marketplace.

Your submitted API will be reviewed by Microsoft before publication. The approval will take a few business days. Let's see how to implement each of the these three steps in more detail.

Creating and Publishing a Web Service for Your Machine Learning Model

In this example, you will use the predictive maintenance model we discussed in Chapter 11. The following two steps are required to deploy your model in production. With these steps you basically create a web service that can be called as an API.

1. Create a Predictive experiment, and
2. Publish your experiment as a web service.

Let's review these steps in detail and see how they apply to your finished model built in Chapter 11. You can start this section by opening the predictive maintenance model in Azure Machine Learning Studio.

■ **Note** The model we will discuss in this section is published as *Predictive Maintenance Model* in the Azure Machine Learning Gallery. You can access the Gallery at `http://gallery.azureml.net/`. Feel free to download this experiment to your workspace in Azure Machine Learning.

Creating Scoring Experiment

To create a scoring experiment, follow these steps in Azure Machine Learning Studio.

1. Run your experiment with the **Run** button at the bottom of Azure Machine Learning Studio.
2. Select the **Train Model** module in the left branch containing the Boosted Decision Tree. This tells the tools that you plan to deploy the Boosted Decision Tree model in production. This step is only necessary if you have several training modules in your experiment.
3. Next, click **Set Up Web Service | Predictive Web Service [Recommended]** at the bottom of Azure Machine Learning Studio. Azure Machine Learning will automatically create a predictive experiment.
4. Your predictive experiment should appear as shown in Figure 13-7. You are now ready to deploy your new predictive experiment in production. To do this, move to the next step.

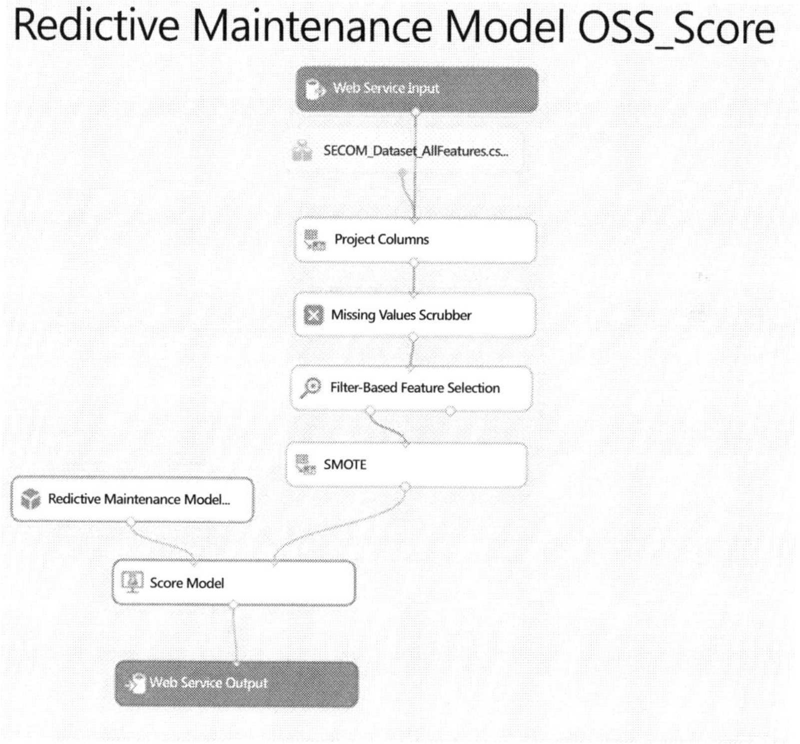

Figure 13-7. *The scoring experiment ready for deployment as a web service*

Publishing Your Experiment as a Web Service

At this point, your model is now ready to be deployed as a web service. To do so, follow these steps.

1. Run your experiment with the **Run** button at the bottom of Azure Machine Learning Studio.
2. Click the **Deploy Web Service** button at the bottom of Azure Machine Learning Studio. The tool will deploy your scored model as a web service. The result should appear as shown in Figure 13-8.

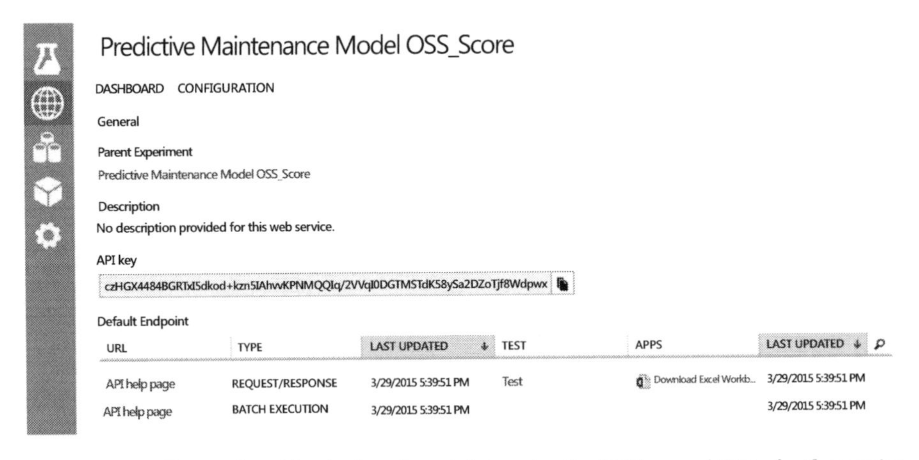

Figure 13-8. *The results of the deployed model showing the API key and URLs for the service*

Obtaining the API Key and the Details of the OData Endpoint

The dashboard in Figure 13-8 shows the API key for your new web service. You will need this to deploy your model in the marketplace. Next, you will need to get the URL of your OData endpoint. To obtain this from the same dashboard, click the first link named **API help page** for the **Request/response** service. This will open the page shown in Figure 13-9. Copy the OData endpoint address.

Request Response API Documentation for Predictive Maintenance Model OSS_Score

Updated: 03/30/2015 00:39

No description provided for this web service.

- Previous version of this API
- Submit a request
- Input Parameters
- Output Parameters
- Sample Code

OData Endpoint Address

https://ussouthcentral.services.azureml.net/odata/workspaces/13d04a39758746ab9bd02f9218087faa/services/55983585051449fb84e459bee86db8cd

Request

Method	Request URI	HTTP Version
POST	https://ussouthcentral.services.azureml.net/workspaces/13d04a39758746ab9bd02f9218087faa/services/55983585051449fb84e459bee86db8cd/execute?api-version=2.0&details=true	HTTP/1.1

*Note: You may omit the **details** parameter from the query string. This would cause **ColumnTypes** to be omitted from the output*

Figure 13-9. *API documentation of the deployed predictive model*

Publishing Your Model as an API in Azure Marketplace

Having deployed your model as a web service, you are ready to publish it to Azure Marketplace. To do this simply follow these steps.

1. Visit Azure Marketplace at `https://datamarket.azure.com/home` and click **Publish** in the top menu on the page.
2. Select **data services** from the menu on the left pane, and enter a name for your new service. See Figure 13-10 for details.

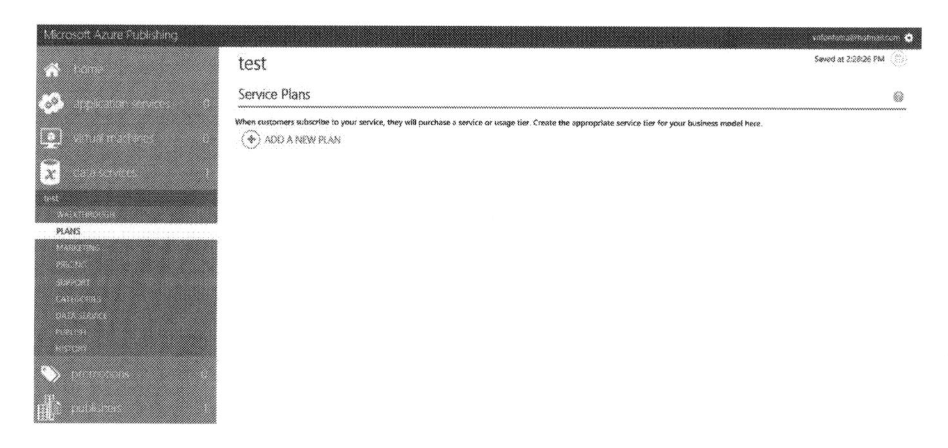

Figure 13-10. *Creating a new data service in Azure Marketplace*

3. Click the link named **Create a new service**.
4. Complete your seller profile. You also need to provide your banking details to receive payments from your customers.
5. Under **Plans**, create pricing plans for your new service. Do browse other services on Azure Marketplace for examples of pricing plans.
6. Provide marketing content in the **Marketing** submenu. This includes descriptions and images.
7. In the pricing submenu, set prices for the countries covered by Azure Marketplace. You can set the prices per country manually. Even better, Azure can automatically calculate prices per country if you click the **autoprice** link.
8. Under the **Data Services** tab, select **Web Services** as your data source.
9. Enter the URL of your OData service (from the last section) in the **Service URL** text field.
10. Choose **Header** as the **Authentication scheme**.
    a. For **Header Name**, enter **Authorization**.
    b. In **Header Value**, enter **Bearer**. Click the space bar, and then paste your API key (the one you obtained in the previous section).
    c. Check the box labeled **This service is OData**.
    d. Now click **Test Connection**. Figure 13-11 shows the completed form. The header name and value are encrypted.

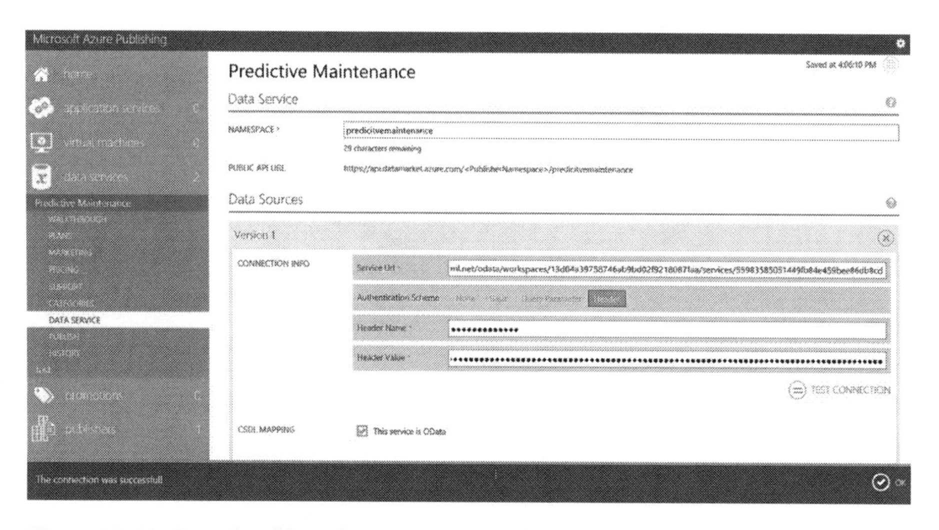

***Figure 13-11.** Completed form for Data Service tab*

11. Be sure to select **Machine Learning** in the **Categories** tab.
12. After testing, select **Publish** and choose **Push to staging**.
    The tool will tell you if there are any issues with your submission.
13. Once you are ready, choose **Request approval to push to production**. Microsoft will review your submission for approval. This can take a few business days.

Congratulations, you have just published your first predictive model in Azure Marketplace. Once approved, you can start earning money from your own predictive model.

Summary

As you saw in this chapter, Azure Marketplace is a great opportunity for you to make money from your predictive models. Azure Marketplace is an online site for buying and selling data and Machine Learning solutions. This chapter introduced this marketplace, showing existing Machine Learning APIs from Microsoft and its partners. You saw how to consume services from Azure Marketplace. Specifically, you reviewed the Recommendations API, a potent recommendation engine available as an API service from Azure Marketplace. You also saw step by step how to sell your own predictive models on Azure Marketplace.

CHAPTER 14

Cortana Analytics

The Cortana Analytics Suite provides companies with a managed big data and advanced analytics suite to transform data into intelligent action. In this chapter, you will learn about the different services in the Cortana Analytics suite, and how the suite can empower your organization to build and compose powerful end-to-end advanced analytics solutions that distil the nuggets of data in your organization into gold. You will also learn about the exciting capabilities delivered by Cortana Analytics.

What Is the Cortana Analytics Suite?

In earlier chapters, you learned about the practice of data science, and how you can use Azure Machine Learning to solve practical problems that can enable your organization to stay one step ahead of your competitors. Some of the practical problems (non-exhaustive) include

- Building customer segmentation models that enable you to better address the needs of your customers.
- Predicting customers that will churn, and figuring out strategies for retaining them before they leave.
- Building customer propensity models to better target customers for marketing campaigns.
- Predicting the yield from a manufacturing plant, and identifying the key factors influencing yield.

To design the architecture for an end-to-end solution that addresses these practical problems, you will need capabilities beyond just machine learning algorithms.

Depending on business requirements, you will need different methods of ingesting data, storing data, and processing data (both real-time and batch processing). You will need to be able to orchestrate, monitor, and manage all the data movement and processing tasks that are necessary to ensure that the analytics machinery runs smoothly on a day-to-day basis. Most importantly, you will need to empower end users to make well-informed and intelligent decisions by enabling them to easily interact with the intelligent system in intuitive ways, and be able to access the insights through rich visualizations and dashboards.

Figure 14-1 shows how to leverage the various capabilities of the Cortana Analytics Suite to turn data into intelligence, and make it actionable for people and automated systems alike.

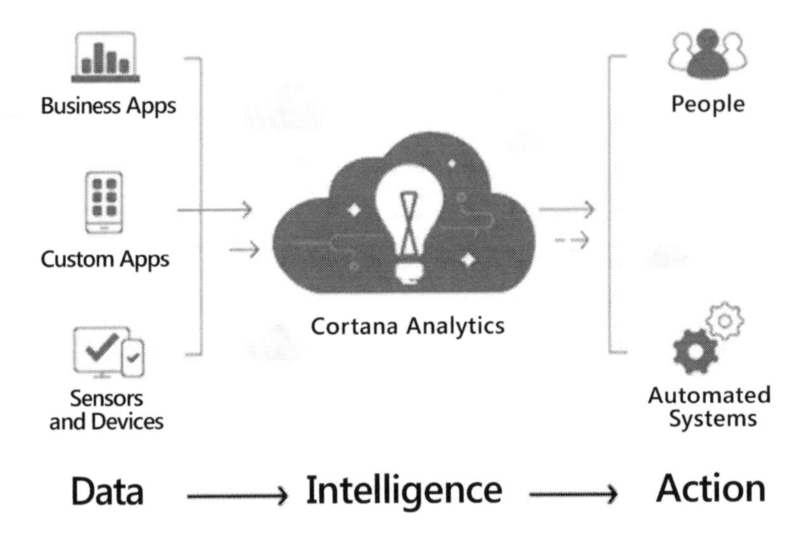

Figure 14-1. *Cortana Analytics Overview (Source: Microsoft,* `http://azure.microsoft.com/blog/2015/07/13/announcing-cortana-analytics-suite-and-new-partner-investments-at-wpc-2015`*)*

■ **Note** To learn more about Cortana Analytics, refer to `http://microsoft.com/cortanaanalytics`.

Capabilities of Cortana Analytics Suite

The Cortana Analytics Suite comprises of the following capabilities:

- Machine Learning and Analytics
- Perceptual Intelligence
- Big Data Stores
- Information Management
- Dashboards and Visualizations

In addition, Cortana Analytics enables you to leverage Cortana as your personal digital assistant. Pre-packaged solutions enable you to jumpstart various business scenarios through APIs offered in the marketplace (such as recommendations, forecasting, churn, face, vision, speech, and text analytics), and templates provided in the Azure Machine Learning Gallery. Within the Cortana Analytics suite, the machine learning and analytics capabilities are provided by Azure Machine Learning, Azure HDInsight, and Azure Stream Analytics. Azure HDInsight and Azure Stream Analytics address the batch processing and real-time processing needs of an organization. Azure Stream Analytics and the event hub enable you to handle data ingestion from a large number of devices/sensors and perform real-time processing of the data. In addition, Azure HDInsight provides a managed Hadoop service to process the data at scale, using various big data technologies (such as Pig, Hive, Spark, etc.). When composed together in an end-to-end solution, these building blocks enable you to build powerful analytic solutions.

Perceptual Intelligence capabilities in the Cortana Analytics suite are provided by face, vision, speech, and text analytics APIs. The face and vision APIs enable you to perform face and object detection and matching. Speech APIs enable you to analyze speech and convert spoken audio to text and/or intent. Text analytics enables you to perform sentiment analysis and extract key phrases from text. For example, you can use the face APIs to figure out the age and gender of customers, automating the verification and identification of faces, as well as grouping of faces. You can leverage these APIs (combined with machine learning and analytics capabilities) to quickly build intelligent applications that provide customers with personalized experiences.

Various big data stores are provided as part of the Cortana Analytics suite. These range from Azure storage and Azure SQL Database to powerful capabilities offered by the Azure Data Lake and elastic data warehouse capabilities offered by Azure SQL Data Warehouse.

To deal with the challenges of information management, the Cortana Analytics suite provides data orchestration capabilities using Azure Data Factory. Azure Data Factory enables you to easily compose data movement and processing tasks. For example, you can leverage Azure Data Factory to move data from various data stores, and process the data (such as aggregating the data) before leveraging Azure Machine Learning web services to perform predictions on the data. Most importantly, it provides you with a single pane of glass for monitoring and managing the entire solution. When building the solution, you will often need to discover the various data sources available to you. Azure Data Catalog provides a managed service for you to find and use relevant data for your projects.

Power BI provides rich visualizations and dashboards that will enable the business stakeholders to get a holistic view of the business, and analyze and understand the insights that are delivered by the various services in the Cortana Analytics suite.

Example Scenario

To illustrate how businesses can leverage the Cortana Analytics Suite, let's extend the mobile operator scenario from Chapter 9. In Chapter 9, you learned how data scientists built a model to identify which mobile customers will churn. After the customer churn model has been built and validated, the model is published as a web service. Figure 14-2 shows how the different services in the Cortana Analytics Suite can be used in a preconfigured solution to predict churn for customers in the telecommunication industry.

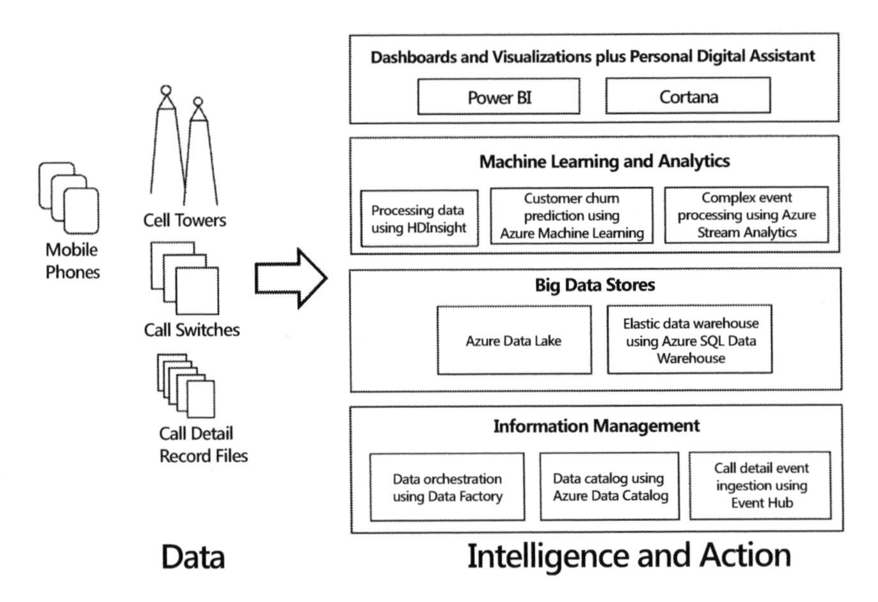

Figure 14-2. *Customer churn solution using the Cortana Analytics Suite*

To effectively figure out whether a mobile customer is likely to churn, the customer churn model needs to leverage features that include customers' profile (age, gender, education, marital status, etc.), as well as customer usage information (how many minutes the customer spends each month, whether they have been paying their bills on time, etc.). To build the end-to-end solution, the mobile operator will need to identify the relevant data sources available in the company. Azure Data Catalog will enable them to easily discover the information that they need as part of the solution. To operationalize this as an end-to-end solution using Azure Machine Learning, the mobile operator needs to preprocess the data so that it can be used as inputs to the published Azure Machine Learning web service.

To compute the customer usage information, most mobile operators rely on the processing of call detail record files (CDR files). Using Azure HDInsight, the mobile operator can process and aggregate the customers' usage at scale, and figure out the time spent by each customer monthly. The aggregated customer information is combined with the customers' profile, so that a complete view of the customers can be created and used as inputs to the published Azure Machine Learning web service, which predicts the customers who are likely to churn. To provide the business stakeholders with a holistic view of the business and key performance indicators (such as total number of customers per segment and number of customers that are likely to churn in the next few months), a Power BI dashboard is created.

In addition, the mobile operator wants to drive a personalized retail experience where customers are provided with recommendations on the type of phones and accessories that best suits their lifestyle, gender, and age group. To build these personalized experiences, the mobile operator leverages the Perceptual Intelligence capabilities provided by the Cortana Analytics Suite. As a customer walks into the retail store and browses a display, the face detection APIs kick into action. Using various salient characteristics of the customer (like age, gender, and composition of the group that is browsing the display), an intelligent kiosk is able to recommend relevant phones and accessories that are applicable to customers of the specific age group, recommend family plans or plans for a group of friends, and much more.

As the demands of the business grow, the mobile operator evaluates how they can leverage big data stores in the Cortana Analytics suite for storing the latest data as well as archived data (data from transactions that occurred more than 18 months ago). Azure Data Lake provides a hyperscale big data store for both structured and unstructured data. To build a highly scalable data warehouse in the cloud to power their business, the mobile operator explores how they can tap on the elastic warehouse capabilities offered by the Azure SQL Data Warehouse.

Summary

In this chapter, you learned about the core capabilities of the Cortana Analytics suite, and how you can leverage them in your organization to turn data into intelligent actions.

Index

A

B

C

D

E

F

G

H

I, J

K

L

M

N

O

P

Q

R

S

T

U

V

W, X, Y, Z

Apress®
THE EXPERT'S VOICE™